ELECTROMAGNETIC WAVES AND HEAT TRANSFER

SENSITIVITIES TO GOVERNING VARIABLES IN EVERYDAY LIFE

ELECTROMAGNETIC WAVES AND HEAT TRANSFER

SENSITIVITIES TO GOVERNING VARIABLES IN EVERYDAY LIFE

by
M. Kemal Atesmen

ASME PRESS

Library of Congress Control Number: 2020947628

Contents

Preface	vii
Acknowledgements	ix
About the Author	xii
Introduction	1
1 Our Sun and Thermal Radiation Distribution Function	9
2 Total Hemispherical Emissivity	27
3 Thermal Radiation Configuration Factor Example	35
4 Vertical Radiant Heater	41
5 Center of Glass Heat Transfer in Vertical Double-Pane Windows	47
6 Thermal Radiation Effects on Human Skin From a Nuclear Explosion	59
7 Exposure to Solar Ultraviolet Radiation On Earth	69
8 Exposure to Ultraviolet Radiation From Fluorescent Lamps	77
9 Radiation Heat Transfer from a Thermos Bottle	83

10 Radiation and Laminar Forced Convection Heat Transfer During a Vehicle's Paint Curing Process 89

11 Sun's Irreplaceable Energy Source Optimization for a PV Collector Panel's Tilt Angle 101

12 Converting Solar Radiation to Thermal Energy With a Glazed Flat Plate Collector 113

13 Human Thermal Comfort 123

14 Energy Balance for a Swimming Pool 133

15 Radiation Heat Transfer in Unsteady State 143

16 Cooling of Steel Spherical Balls by Natural Convection and Radiation Heat Transfer in Unsteady State 149

17 Baking Pizza by Infrared Radiation and Conduction Heat Transfers and Mass Transfer in Unsteady State 157

18 Microwave Radiation in Cooking Ovens 165

19 Electromagnetic Radiation Used in Medical Imaging 171

20 Storing Liquid Oxygen in a Cryogenic Spherical Container 175

21 Storing Liquid Oxygen In a Cryogenic Spherical Container With a Radiation Shield 181

22 Temperatures of Different Surfaces Exposed to Incident Extraterrestrial Radiation at Several Locations 185

23 Cell Phone's Radio Frequency Electromagnetic Radiation Effects on Human Brain Tissue 193

24 Cooling a Radar's Electronic Board 199

References 203

Index 205

Preface

Electromagnetic waves generate radiation energy and they play very significant roles in our lives. Electromagnetic waves are studied in almost every scientific field from astronomy, agriculture, chemistry, medicine to physics. This book focuses on heat transfer aspects of electromagnetic waves. There are twenty-four chapters in this book with their solutions to heat transfer from electromagnetic waves' radiation energy with different uses and problems related to our lives. Each problem solution also investigates the sensitivity of critical independent variables to governing dependent variables.

For example, our Sun's electromagnetic waves that reach our Earth make this place livable. We are lucky to have our Sun. It is a yellow dwarf star with an average surface temperature of 5800 K and it acts like an ideal thermal radiator. Some of our Sun's electromagnetic waves reaching us on Earth can be harmful to us due to their high energy levels and others with low energy levels can be very beneficial to us. Radiation energy received from a body with a surface temperature greater than zero Kelvin can be determined by summing all the thermal radiation contributions over the entire wavelength spectrum. Radiation energy exchange between two bodies can be calculated with ease if they are perfect absorbers and emitters and their surface temperatures are known and also if the configuration factor between these two surfaces is determined. If a body is not a perfect absorber and emitter, then we have to determine emissivity of a surface as a function of surface temperature, wavelength and direction of radiation emission.

About 95 % of total solar radiation reaching Earth's surface comprises of visible and infrared regions of the electromagnetic wavelength spectrum. Remaining 5 % of total solar radiation reaching Earth's surface comprises of ultraviolet (UV) region. Exposure to solar UV radiation can have cumulative adverse effects on human health. Solar radiation that strikes our windows, swimming pools, roofs, and photo-voltaic collectors provide us free and very useful energy.

When we are cold on our patios or outside of a restaurant, a thermal radiation heater in the infrared electromagnetic wave region can warm us. In a nuclear explosion, the fireball emits large amounts of ultraviolet, visible and infrared electromagnetic waves in the first few seconds that can devastate everything around point of burst.

Fluorescent lamps for general lighting come in different powers, sizes and phosphors to emit visible radiation energy. Our thermos bottles keep our coffees hot or our soft drinks cold mainly by reflecting electromagnetic waves between two surfaces.

In order to achieve the best paint performance for our vehicles, each layer of paint has to be cured in a bake oven according to paint supplier's bake temperature versus time specifications. Accurate control of bake temperature versus time specification is critical to the performance of a coating both physically and cosmetically. This accurate temperature control in a paint curing bake oven is obtained by laminar forced convection heat transfer and by infrared radiation heat lamps.

Even during baking a pizza, infrared electromagnetic waves from hot ceiling bricks of a hemispherical dome of a wood fired oven and by conduction heat transfer from hot bricks covering the floor are used to have a perfect pizza crust and its toppings. Microwave radiation cooking ovens are an irreplaceable component of a kitchen. Microwaves, in the electromagnetic wave spectrum, fall between the infrared radiation at the high frequency end, i.e. 300 GHz at 1 mm wavelength, and radio waves at the low frequency end, i.e. 300 MHz at 1 m wavelength.

Different bands of microwaves are used in different technologies ranging from our cell phones, garage door openers, point-to-point wireless communication links, satellite and spacecraft communications to advanced weather forecasting by providing global view of weather patterns and surface temperatures. Cosmic microwave background radiation fills the entire universe and it provides us knowledge about the beginnings of our universe, namely about the Big Bang. theory.

A radar transmits from its transmitter short pulses of electromagnetic waves at radio and microwave frequencies, i.e. 30 kHz to 100 GHz, which may be reflected by objects on their path. Reflected electromagnetic waves come back to the receiving portion of a radar system to provide information about the shape and speed of the reflecting object such as airplanes, boats, missiles, speeding cars, etc.

Acknowledgements

Over 40 years of engineering, engineering management and project management in the global arena covering automotive, computer, data communication, and offshore oil industries were accomplished by exceptional support from my wife, Zeynep, and my family members. Some years I was away from home more than six months out of a year trying to tackle challenging engineering tasks.

I would like to dedicate this book to all engineering project teams' members with whom I had the pleasure of working together over the years, with enthusiasm, with imagination and with determination. Over these years my engineering project teams' members kept coming back to work along with me without any reservations.

M. Kemal Atesmen
Santa Barbara, California

About the Author

M. Kemal Atesmen completed his high school studies at Robert Academy in Istanbul Turkey in 1961. He received his B. Sc. degree from Case Western Reserve University, his M. Sc. degree from Stanford University, and his Ph. D. degree from Colorado State University, all in mechanical engineering. He is a life member of ASME. He initially pursued an academic and an industrial career in parallel and became an associate professor in mechanical engineering before dedicating his professional life to international engineering management and engineering project management for thirty-three years. He helped many young engineers in the international arena to bridge the gap between college and professional life in automotive, computer component, data communication, and offshore oil industries.

He published eight books, sixteen technical papers, and has four patents. His books are "Global Engineering Project Management", CRC Press, 2008, "Everyday Heat Transfer Problems – Sensitivities to Governing Variables", ASME Press, 2009, "Understanding the World Around through Simple Mathematics", Infinity Publishing, 2011, "A Journey Through Life, Wilson Printing, 2013, "Project Management Case Studies and Lessons Learned", CRC Press, 2015, "Process Control Techniques for High Volume Production", CRC Press, 2016, "Engineering Management in a Global Environment: Guidelines and Procedures", CRC Press, 2017, "Case Studies in Fluid Mechanics with Sensitivities to Governing Variables", Wiley and ASME Press, 2018, and "Nothing Is Fixed, Everything Changes", Archway Publishing, 2018.

Introduction

In this book effects of electromagnetic waves that play significant roles in our lives through radiation heat transfer are investigated in twenty-four chapters. The approach to a problem's solution in a chapter starts with an overview of electromagnetic waves and basic laws of radiation heat transfer, mass transfer and fluid mechanics. Then simplifying engineering assumptions are discussed and governing equations, dependent and independent variables are identified. In some cases, where solutions to basic equations are not possible, past experimental studies are utilized. Solutions to governing equations are described and presented graphically. Also, analyses are extended to sensitivities of dependent variables to independent variables within the region of interest.

Chapter 1 explains the radiant power of our Sun that reaches Earth and other planets in our solar system. Our Sun is a yellow dwarf star which emits radiation equally in all directions like a diffuse emitter and over all wavelengths of electromagnetic waves, namely an ideal thermal radiator. Our Sun's electromagnetic radiation spectrum characteristics are detailed. Distributions of incoming, outgoing and absorbed radiation power from our Sun to Earth are discussed. An overview of radiation power released by Earth's surface into the atmosphere is also discussed. Thermal radiation of black bodies which are perfect electromagnetic radiation absorbers and emitters are discussed. A black body emits thermal radiation due to its surface temperature at different wavelengths and frequencies. Total thermal radiation from the surface of a black body with a surface temperature greater than zero Kelvin is determined by summing all the thermal radiation contributions over the entire wavelength spectrum using Plank's distribution function.

Chapter 2 deals with total hemispherical emissivity which is the emissivity of a surface that is averaged over all wavelengths and hemispherical directions and which is only a function of surface temperature.

Chapter 3 explains the net exchange of thermal radiation between two black surfaces. We only have to know the configuration factor or view factor between these two black surfaces, their surface temperatures and the surface areas of these two surfaces. The configuration factor or view factor is defined as the fraction of thermal radiation that leaves a surface and intercepted by another surface. As an example, this chapter analyzes the configuration factor from a small diameter cylinder that is perpendicular to the center of a disk.

In Chapter 4, effectiveness of a rectangular and vertical thermal radiation heater to a rectangular and vertical elemental surface that is located at the base of the heater at different locations is analyzed by determining the configuration factor between the elemental surface and the vertical thermal heater.

Chapter 5 investigates center of glass heat transfer in vertical double pane windows under steady-state heat transfer conditions. It is assumed that vertical double panes are parallel, air tight and with no moisture condensation. Heat flow is perpendicular to glass pane surfaces and it is one-dimensional. Also, solar radiation strikes the panes at near normal incidence. In the present analysis, it will be assumed that the outer and the inner surfaces of the glass layer has the same temperature and they are isothermal, since the glass layers are thin. Radiation heat transfer ignores gas absorption between panes and convective heat transfer assumes high aspect ratio panes.

In Chapter 6, thermal energy that radiates from the surface of a fireball resulting in a nuclear explosion is investigated. Only nuclear explosions (bursts) occurring in air above the surface of the Earth and below five kilometers altitude are considered. Approximately 35% of energy released from a nuclear explosion is in the form of thermal radiation. Surface of a nuclear explosion fireball emits large amounts of ultraviolet, visible and infrared rays in the first few seconds after a nuclear explosion. Published experimental results of thermal radiation power of a nuclear fireball versus time after explosion are used.

About 95 % of total solar radiation reaching Earth's surface comprises of visible (380 to 750 nanometers, nm wavelengths) and infrared (700 to 1×106 nm wavelengths) regions of the electromagnetic radiation spectrum. Remaining 5 % of total solar radiation reaching Earth's surface comprises of ultraviolet (UV) (100 to 400 nm wavelengths) region of the electromagnetic radiation spectrum.

Chapter 7 analyzes the UV radiation on Earth's surface that effects the human skin, namely UV induced erythema on the human skin, using spectrally resolved spectroradiometer data taken at a tropical location and at a location close to the polar circle. UV radiation data is reduced to standard UV index (UVI) formulation as defined by World Health Organization and by ISO Standard 17166.

Chapter 8 investigates exposure to UV radiation from fluorescent lamps. Fluorescent lamps for general lighting come in different powers, sizes and phosphors to emit visible radiant energy. A plastic cover called diffuser that is placed over the lamp effectively eliminates all ultraviolet radiation from the lamp. However, if there is no cover over the lamp and the glass envelope of the lamp is defective and has metal impurities in it, the ultraviolet radiation (UVR) emission from the lamp can cause significant cumulative damage to human eyes and skin.

In Chapter 9, heat transfer from a thermos bottle is detailed. Dewar invented the thermos bottle at the beginning of the twentieth century which is simply a stainless steel inner bottle covered by another stainless steel outer bottle which are separated by vacuum. The outer wall of the inner bottle and the inner wall of the outer bottle are coated with silver in order to minimize radiation heat transfer between the fluid inside the thermos bottle and the outside environment. Two silver coated walls facing each other reflect electromagnetic waves between themselves in the infrared region and minimize radiation heat transfer from the medium inside the inner bottle to the outside environment or vice versa. There is a minor amount of heat loss due to conduction and convection from the lid of the thermos bottle where a small portion of the inner surface of the stainless steel inner bottle contacts the lid.

Chapter 10 investigates radiation and laminar forced convection heat transfer during a vehicle's exterior pain curing process. Painting a vehicle involves several layers such as electrocoating that is applied to the virgin metal body, primer, base coat, clear coat, sealants and structural adhesives. In order to achieve the best paint performance, each layer of paint has to be cured in a bake oven according to paint supplier's bake temperature versus time specifications. Accurate control of bake temperature versus time specifications is critical to the performance of a coating both physically and cosmetically. During the baking process, complete cross linking of the paint has to be achieved. Chapter 10 investigates a paint curing bake oven where heat is supplied to a particular vehicle's painted body by hot air flow, i.e. by laminar forced convection heat transfer, and by infrared radiation heat transfer lamps.

Chapter 11 optimizes a photovoltaic collector (PV) panel's tilt angle for Sun's energy. By the year 2040, energy coming from our Sun increases eight-fold to 9.66% of World energy consumption by replacing coal and oil. This bright future in solar radiation energy consumption makes everyone eager to learn about PV panel system design and optimization of solar radiation energy conversion into electrical energy. This optimization process starts with the knowledge of solar radiation energy available at a particular location and at a particular time on Earth. For Example, a PV collector panel in San Diego, CA at the latitude of 32.73 degrees North, that is facing South at an optimum tilt angle from horizontal, will receive a total radiant energy, i.e. sum of direct beam, diffuse and ground reflected energy irradiation, of 2,091 kWh/(m^2-year). A similar PV collector panel facing South in Anchorage, AK at the latitude of 61.17 degrees North, at its optimum tilt angle, will only receive a total radiant energy of 1,086 kWh/(m^2-year).

Chapter 12 details conversion of solar radiation to thermal energy with a glazed flat plate collector. A detailed heat transfer analysis for converting solar energy into useful thermal energy for a glazed flat plate solar radiation collector can be very involved and complicated. However, a simple heat transfer analysis between the incident solar radiation and the useful thermal energy gained by the transfer fluid can provide us with a good measure of a glazed flat plate collector's performance. The function of a solar radiation collector is to absorb solar radiation energy by means of a high absorptivity surface and convert it into thermal energy. The thermal energy is then transferred from the absorber surface to flowing fluid by means of high thermal conductivity tubes. The collector is protected from convective losses to the environment with a thin film coated glass cover that is highly transparent to incoming radiation, but prohibits short wavelength radiation leaving the collector due to its very low emissivity. Insulating material is placed along sides and back of the collector in order to minimize convective heat losses to the environment.

Humans are very sensitive to changes in environmental temperature and humidity because our internal body temperature has to be kept at a narrow temperature range of 37 ± 0.5 C. When our internal body temperature rises above 37.5 C, our blood flow is increased to our skin in order to increase the heat transfer from our skin to the environment. At elevated internal body temperature sweating also starts and cools our skin which increases the heat transfer from the body core to our skin. When our internal body temperature falls below 36.5 C, our blood flow is decreased to our skin in order to decrease the heat transfer from our skin to the environment. Also, when our internal body temperature falls below 36.5 C, muscle shivering can start in order to increase the internal metabolic heat generation. Thermal energy balance between a human body and the environment under steady-state conditions is investigated in Chapter 13.

Swimming pools use solar collectors or natural gas heaters to maintain comfortable temperatures that are desired, i.e. between 20 C to 30 C. UV stabilized transparent plastic or bubble type covers are used during night to minimize losses due to evaporation heat transfer and convection heat transfer from pool water's surface. Since water in a pool has a very large thermal mass, water's temperature changes slowly. Therefore, in Chapter 14, steady-state heat transfer conditions are analyzed. Also monthly averaged solar radiation absorbed by water, relative humidity of air, latent heat of evaporation and diffusion coefficient of water in air are used. Conduction heat transfer from pool walls to earth are neglected, since this mode of heat transfer is less than 2% as compared to major heat transfer mechanisms, such as incident solar radiation, evaporation from surface of pool water, convection losses from surface of pool water and infrared radiation losses from surface of pool water.

Chapter 15 analyzes radiant heating in unsteady-state of spherical balls with different radii that are placed at the center of a spherical oven. The oven has a temperature controlled inner surface for infrared radiant heating. The spherical ball is heated by radiation in a vacuum environment, namely there is no convection heat transfer at the surface of the spherical ball.

In Chapter 16, steel spherical balls at different radii that were heated to 400 C (673 K) by infrared radiation in a vacuum environment in Chapter 15 will be cooled down slowly by natural convection and by infrared radiation heat transfer in unsteady-state down to a temperature of 50 C (323 K). A hot spherical ball will be placed at the center of a large air chamber in which chamber air and chamber walls are kept at a constant temperature at 25 C (298 K). This kind of slow cooling down of hot metals are done in order to relieve stresses that are generated in the metal structures during their processing. It is assumed that the temperature inside the spherical ball is uniform at any instant of time during the unsteady-state cooling process.

Chapter 17 analyzes the baking of a pizza by infrared radiation and by conduction heat transfers and mass transfer in unsteady-state. The pizza is heated by infrared radiation heat transfer from hot ceiling bricks of a hemispherical dome of a wood fired oven and by conduction heat transfer from hot bricks covering the floor which are contacting the bottom surface of a pizza crust. During the baking process, pizza will also lose heat by mass transfer, namely by evaporation, from pizza's top surface to oven's air. All heat transfer mechanisms acting on the pizza are in unsteady-state from the moment the pizza is placed on the floor of the oven. Objects of baking a good pizza is not to burn the bottom of the crust and have a crispy bottom and also cook the dough and toppings to an exact water content at the end of baking process.

Chapter 18 investigates the microwave radiation in cooking ovens. In a microwave oven, electromagnetic waves generated by a magnetron at the frequency of 2.45 GHz, a wavelength of 12.2 cm, enter an oven's cavity and penetrate the food which is placed at the center of a rotating turntable. Microwaves pass through most food containers made out of glass, paper and plastic without heating them. Part of the food molecules such as water or fat carry a slight positive charge while the other part of these food molecules carry a slight negative charge. This charge imbalance in food molecules colliding with the changing electrical field of the electromagnetic microwaves cause dipole rotations. These molecular dipole rotations generate friction and therefore heat between molecules. Food heats up fast and volumetrically. However, very complex interactions between microwaves and uneven food properties such as dielectric and thermal properties, produce non-uniform heating.

Electromagnetic radiation that are used in medical diagnosis are discussed in Chapter 19. There are two categories of electromagnetic radiation that are used extensively in medical diagnosis. One category is ionizing radiation which is emitted as gamma rays or x-rays during the imaging processes such as radiography (x-ray), computer tomography (CT scan uses x-rays) and nuclear medicine (uses gamma rays). The other category is non-ionizing electromagnetic radiation which is emitted as radio waves or sound waves during the imaging processes such as magnetic resonance imaging (MR) which uses radio frequency waves and ultrasound imaging which uses sound waves. Typical effective radiation doses received from different medical diagnosis are given.

Chapter 20 investigates the radiation heat transfer from a cryogenic spherical container for storing liquid oxygen. Liquid oxygen is used widely in industrial and medical fields and also in spacecraft rocket propellants. Liquid oxygen storage is less bulky and less costly than the equivalent capacity of high pressure gas oxygen storage. The freezing point of liquid oxygen in one atmosphere of pressure is –219 C (54 K). The boiling point of liquid oxygen in one atmosphere of pressure is –183 C (90 K). This chapter investigates the minimization of radiation heat transfer from a cryogenic spherical container for storing liquid oxygen.

Chapter 21 investigates the radiation heat transfer from a cryogenic spherical container with a radiation shield. Storing liquid oxygen in a cryogenic spherical container with a concentric spherical outer shell was analyzed in Chapter 20. In this chapter, a concentric spherical radiation shield with low emissivity surfaces is inserted in between the cryogenic spherical container and the outer spherical shell in order to minimize radiation heat losses from the liquid oxygen.

Chapter 22 analyzes steady-state temperatures of three different surfaces exposed to incident extraterrestrial radiation at four different locations. Three type of horizontal surfaces whose back sides are well insulated are analyzed. These surfaces have white paint or black paint on them or made out of red brick. They are diffuse and gray surfaces and their absorptivities, α, and emissivities, ε, are given below:

White paint on metallic surface:	$\alpha = 0.24$,	$\varepsilon = 0.96$
Black paint on metallic surface:	$\alpha = 0.97$,	$\varepsilon = 0.97$
Red brick surface:	$\alpha = 0.63$,	$\varepsilon = 0.93$

The locations are Anchorage, Alaska at 61.17 degrees latitude north, Seattle, Washington at 47.61 degrees latitude north, Sacramento, California at 38.58 degrees latitude north and San Diego, California at 32.73 degrees latitude north. All four locations are very close to the sea level, namely altitudes vary between 9 meters and 35 meters. So altitude effects are neglected in the calculations.

Chapter 23 analyzes cell phone's radio frequency electromagnetic radiation effects on human brain tissue. Conservation of energy at any instant in time to a brain tissue that is just inside the human skull located behind the ear is analyzed with a one-dimensional transient heat transfer model by neglecting the heat gained due to metabolism. Cell phones and digital cellular networks can produce high enough non-ionizing radio frequency electromagnetic radiation energy during long exposures, i.e. more than 1 hour, that can increase temperature of brain tissues high enough to affect the human nervous system. International guidelines allow a localized maximum exposure which is specified as 10 W/kg specific absorption rate (SAR), averaged over a 10-gram tissue for non-ionizing radio frequency electromagnetic radiation energy. Radio frequency radiation includes radio waves and microwaves. In the new fifth generation (5G) wireless technology for cell phones and for digital cellular networks, electromagnetic radiation waves operate between 30 and 300 GHz frequency which can produce high radiation energy levels and damage human tissues.

A radar, an acronym for **RA**dio **D**etecting **A**nd **R**anging, transmits from its transmitter short pulses of electromagnetic waves at radio and microwave frequencies, i.e. 30 kHz to 100 GHz, which may be reflected by objects on their path. Chapter 24 investigates temperature control of two electronic boards in a radar system using fully developed air flow in a rectangular duct. Air cooling of the electronic boards starts at the fully developed velocity profile region where the electronic boards' surfaces start. Signal to noise ratio of a radar system enhances linearly with transmitted power and inversely with system temperature. Heat generating components of a radar system are the power supply, transmitter, receiver, digital processor and displays. Earlier radar system components and electronic boards had their temperature controlled by using air flow as a coolant. However, as transmitter powers increased, water and ethylene glycol were used for temperature control of electronic boards. Today's high power radars require further enhanced cooling such as two-phase flow systems which combine effects of forced convection cooling and evaporative cooling.

1

Our Sun and Thermal Radiation Distribution Function

We are lucky to have our Sun. It is a yellow dwarf star and at its surface, called the photosphere, it has an average temperature of 5800 K and it acts like an ideal thermal radiator. The Sun is at a mean distance of 1.496×10^{11} meters away from Earth. Sun's gravity holds our solar system together. Also, we are very lucky to have an ideal Earth's surface and an atmosphere around her that balance out the incoming total radiation power with the outgoing radiation power for us to be able to live our lives on this beautiful planet in stable temperatures.

Let us assume that our Sun emits radiation equally in all directions, namely a diffuse emitter, and over all wavelengths of electromagnetic waves like an ideal thermal radiator. This ideal thermal radiator is called a blackbody radiator which absorbs all incident thermal radiation. The sum of all radiation power emitted over all wavelengths by a blackbody is given by the famous Stefan-Boltzmann Equation 1-1, Ref. [33].

$$E_b = \sigma \times T^4 \qquad \text{1-1}$$

where σ is called the Stefan-Boltzmann constant and it has a value of $5.67\text{x}10^{-8}$ $W/(m^2 - K^4)$. T is the surface temperature of the blackbody in degrees Kelvin. A blackbody emits radiation if its temperature is greater than 0 K, namely greater than absolute zero temperature. The radiation power per unit surface area that is emitted by our Sun can be calculated by using Equation 1-1 and the Sun's surface temperature. The result is shown in Equation 1-2.

$$E_{b\text{-}Sun} = 6.42\times10^7 \text{ W/m}^2 \qquad \text{1-2}$$

The mean radius of our Sun, R_{Sun}, is 6.94×10^8 meters. Assuming that our Sun is a perfect sphere, the total radiation power that is emitted by our Sun is calculated from Equation 1-2 by multiplying it by $4\pi R_{Sun}^2$ and the result is shown in Equation 1-3.

$$E_{b\text{-}Sun\text{-}Total} = 3.91\times10^{26}\ \text{W} \qquad 1\text{-}3$$

Total radiation power emitted by our Sun diffuses equally in all directions. Electromagnetic waves travel freely through near vacuum of space between our Sun and the Earth. The radiation power that reaches our Earth's outer atmosphere after traveling a distance of one astronomical unit (AU), namely 1.496x10^{11} meters which is the radius of the sphere, R_{AU}, that our Sun's total radiation power diffuses to reach the Earth. Equation 1-3 can be divided by $4\pi R_{AU}^2$ to determine the radiation power per unit surface area that reaches the Earth's outer atmosphere from our Sun assuming normal incidence. The result which is called the solar constant is given in Equation 1-4.

$$E_{Sun\text{-}Earth} = 1.389\ \text{W/m}^2 \qquad 1\text{-}4$$

Most communication satellites are located at the geostationary orbit which is 35,700 meters above the equator at sea level. These satellites utilize solar cells to operate due to the radiation power that reaches from our Sun to outer edges of Earth's atmosphere. This radiation power from our Sun that reaches outer edges of Earth's atmosphere fluctuates by about 7% during a year, namely from a high of 1,415 W/m^2 around 3 January (at the shortest distance to the Sun, called perihelion) to 1,325 W/m^2 around 4 July (at the farthest distance to the Sun, called aphelion). However, our Sun's total radiation power output changes by time due to solar sunspot activities. Satellite observations throughout several years show that the value that is given by Equation 1-4 is not a constant.

The current correct value for our Sun's total radiation power that reaches the Earth's outer atmosphere is 1,360.8 ± 0.5 W/m^2. See Ref. [22]. The value that was calculated in Equation 1-4 is 2.1% higher than the value of satellite observations, namely 1,360.8 ± 0.5 W/m^2. In this chapter, let us use this total radiation power value determined by satellite observations. The Earth only intercepts ¼ of the total radiation power, namely 340 W/m^2, that reaches from our Sun similar to a flat disk with a mean radius of $R_{Earth} = 6.37\times10^6$ m. The effective flat disk area on Earth that receives our Sun's radiation power is πR_{Earth}^2. Therefore Earth's surface, as she rotates around her axis, receives only $\left(\dfrac{\pi R_{Earth}^2}{4\pi R_{Earth}^2}\right) = \dfrac{1}{4}$ of incoming radiation power from our Sun. If we assume that the Earth behaves like a blackbody with no atmosphere, then the temperature on Earth would be as given in Equation 1-5:

$$T_{Earth} = (340/\sigma)^{1/4} = 278K = 5C \qquad 1\text{-}5$$

Total radiation power from our Sun reaching other planets in our solar system and their temperature are shown in Table 1-1 assuming that these planets have no atmosphere and they behave like blackbodies. As expected, a planet's average surface temperature decreases with increasing distance from our Sun. However, Venus' actual average surface temperature is much higher than our calculations (see sixth and seventh columns in Table 1-1) which is an exception due to its dense atmosphere that causes a strong greenhouse effect. Also, Mercury's actual average surface temperature is much lower than our calculations which is caused by its slow rotation around its axis.

Electromagnetic radiation in space consists of various regions of wavelengths and therefore frequencies and they all travel at the speed of light, $3.00x10^8$ m/s, in vacuum, namely (frequency = speed of light / wavelength). Regions of electromagnetic radiation are presented in Table 1-2. There are no sharp boundaries between these regions. They overlap their neighbors. We use gamma rays and x-rays that are utilized in medical and security areas. Our skins experience the ultraviolet region from our sun all throughout our lives. Our eyes, namely our optical nerves, are sensitive to the visible region, see Table 1-2b, of the electromagnetic radiation spectrum. Almost half the radiation emitted by our Sun is in the visible region of the spectrum. We perceive the thermal infrared radiation as heat. We use microwaves for communication and for cooking. We use radio waves for broadcasting and for listening into space. We, as humans, use every region of electromagnetic radiation shown in Table 1-2 one way or another.

Table 1.1 Radiation power reaching other planets in our solar system from our Sun and their blackbody surface temperatures with no atmospheres compared with actual average surface temperatures provided in Ref. [23]

Planet	Mean Distance From our Sun, m	Mean Distance From our Sun, AU	Total Radiation Power Reaching From our Sun, W/m^2	Effective Radiation Power Reaching From our Sun, W/m^2	Planet's Black-Body Average Surface Temperature with No Atmosphere, K (C)	Planet's Actual Average Surface Temperature from Ref [23], K, (C)
Mercury	5.81×10^{10}	0.39	9,121	2,280	448 (175)	701 (428)
Venus	1.08×10^{11}	0.72	2,635	659	328 (55)	742 (469)
Earth	1.496×10^{11}	1.00	1,360	340	278 (5)	288 (15)
Mars	2.28×10^{11}	1.52	591	148	226 (-47)	243 (-30)
Jupiter	7.79×10^{11}	5.19	50.7	12.7	122 (-151)	164 (-109)
Saturn	1.43×10^{12}	9.51	15.1	3.8	90 (-183)	133 (-140)
Uranus	2.85×10^{12}	19.0	3.8	0.95	64 (-209)	77 (-196)
Neptune	$4.50x101^2$	30.0	1.5	0.38	51 (-272)	71 (-202)

Table 1-2a Electromagnetic radiation spectrum

Radiation Region	Minimum Wavelength	Maximum Frequency Hz	Maximum Wavel ength μ.m	Minimum Frequency Hz
Gamma Rays	-	-	1×10^{-5}	3×10^{19}
X-Rays	1×10^{-5}	3×10^{19}	1×10^{-2}	3×10^{16}
Ultraviolet Rays Rays	1×10^{-2}	3×10^{16}	0.4	7.5×10^{14}
Visible Rays	0.38	7.89×10^{14}	0.75	4×10^{14}
Infrared Rays	0.70	4.29×10^{14}	1,000	3×10^{11}
Microwave Rays	1,000	3×10^{11}	3×10^{5}	1×10^{9}
Radio wave Rays	1×10^{5}	3×10^{9}	1×10^{14}	3

Table 1-2b Visible region radiation spectrum

Visible Region	Minimum Wavelength μ.m	Maximum Frequency Hz	Maximum Wavelength μ.m	Minimum Frequency Hz
Violet	0.38	7.89×10^{14}	0.45	6.67×10^{14}
Blue	0.45	6.67×10^{14}	0.495	6.06×10^{14}
Green	0.495	6.06×10^{14}	0.57	5.26×10^{14}
Yellow	0.57	5.26×10^{14}	0.59	5.08×10^{14}
Orange	0.59	5.08×10^{14}	0.62	4.84×10^{14}
Red	0.62	4.84×10^{14}	0.75	4.00×10^{14}

Earth's surface and the atmosphere around her that balance out the incoming total radiation power with the outgoing radiation power have been generating stable temperatures for centuries. In 2018 average surface temperature on Earth was about 14.9 C. However, Earth's surface temperature on the average have increased about 1.5 C during the 20th century due to increased demand on fossil fuels which increased the greenhouse effect. Earth's surface temperature will keep going up due to increasing greenhouse effect. We, as humans, have to stop this increase in Earth's surface temperature by relying more and more on renewable energy sources such as our Sun, wind and waves. We have to control the delicate balance of total radiation power on our Earth's surface.

Table 1-3 Distribution of incoming radiation power from our Sun

	Incoming Radiation Power From Our Sun, W/m^2	% Of Total Incoming Radiation Power From Our Sun
Absorbed By Earth's Surface	163	48
Absorbed By Earth's Atmosphere	78	23
Reflected Into Space From Earth's Surface And Atmosphere	99	29
TOTAL	340	100

We have calculated earlier that the Earth outside her atmosphere receives 340 Watts per square meter of radiation power from our Sun. Estimates of how the radiation power coming from our Sun is shared on Earth's surface and atmosphere are shown in Table 1-3.

Estimates of radiation power outgoing into space from Earth's surface and atmosphere are shown in Table 1-4.

Estimates of radiation power absorbed by Earth's surface from our Sun and from Earth's atmosphere are shown in Table 1-5. Most of the radiation power that is absorbed by Earth's surface comes from the atmosphere in the infrared region. Human disturbance of this greenhouse effect keeps increasing in an alarming rate.

Table 1-4 Outgoing radiation power into space from Earth's surface and atmosphere

	Outgoing Radiation Power Into Space, W/m^2	% Of Total Outgoing Radiation Power Into Space
Reflected Into Space From Earth's Surface And Atmosphere	99	29
Infrared Radiation Going Into Space From Earth's Atmosphere	241	71
TOTAL	340	100

Table 1-5 Distribution of total radiation power absorbed by Earth's surface

	Radiation Power Absorbed By Earth's Surface, W/m^2	% Of Total Radiation Power Absorbed By Earth's Surface
Incoming Radiation Power From Our Sun Absorbed By Earth's Surface	163	32.4
Greenhouse Effect From Earth's Atmosphere Absorbed By Earth's Surface	340	67.6
TOTAL	503	100

Estimates of total radiation power released by Earth's surface into her atmosphere are shown in Table 1-6. The Earth's surface releases most of her radiation energy into the atmosphere in the infrared region, namely thermal radiation.

Our Sun, a yellow dwarf star, is estimated to have a life of five billion more years. However, human causes increase the magnitude of radiation power from the atmospheric greenhouse effect day after day which will unfortunately diminish life on our beautiful Earth.

Next, we will deal with thermal radiation of blackbodies which are perfect electromagnetic radiation absorbers and emitters. A blackbody emits thermal radiation due to its surface temperature at different wavelengths and frequencies. Total thermal radiation from the surface of a blackbody with a surface temperature, T, greater than zero Kelvin can be determined by summing all the

Table 1-6 Radiation power released by Earth's surface into atmosphere

	Radiation Power Released By Earth's Surface Into Atmosphere, W/m^2	% Of Total Radiation Power Released By Earth's Surface Into Atmosphere
Radiation Power Released By Earth's Surface Into Atmosphere Due To Convection And Evaporation	105	20.9
Radiation Power Released By Earth's Surface Into Atmosphere in the Infrared Region	398	79.1
TOTAL	503	100

thermal radiation contributions over the entire wavelength spectrum as shown in Equation 1-6.

$$E_b = \sigma \times T^4 = \int_0^\infty E_{b\lambda} \times d\lambda \qquad 1\text{-}6$$

where E_b is the total radiation power emitted by a blackbody in W/m^2, a is the Stefan-Boltzmann constant with the value of 5.67x10^{-8} W/(m^2-K^4) and λ is the wavelength in μm of the electromagnetic waves which range from zero to infinity. $E_{b\lambda}$ is the distribution of thermal radiation emitted by a blackbody at a certain wavelength and at a certain temperature in W/(m^2-μm). There have been many attempts to define $E_{b\lambda}$ by several scientists like Rayleigh, Wien and Planck. However Planck's distribution function which is the one given in Equation 1-7, complies with all necessary physical boundary conditions and fits all experimental data.

$$E_{b\lambda} = (C_1/\lambda^{-5})/[exp[(C_2/(\lambda \times \mathrm{T}))-1] \qquad 1\text{-}7$$

where C_1 equals to 3.74x10^8 W-μm^4/m^2 and C_2 equals to 1.44x10^4 μm-K. Derivation of these values can be found in reference Ref [33]. Four boundary conditions for the thermal radiation distribution function given in Equation 1-7 are provided in Equations 1-8.

$$\text{At } T = 0 \qquad E_{b\lambda} = 0 \qquad 1\text{-}8a$$

$$\text{At } T \rightarrow \infty \qquad E_{b\lambda} \rightarrow \infty \qquad 1\text{-}8b$$

$$\text{At } \lambda = 0 \qquad E_{b\lambda} = 0 \qquad 1\text{-}8c$$

$$\text{As } \lambda \rightarrow \infty \qquad E_{b\lambda} \rightarrow 0 \qquad 1\text{-}8d$$

Now let us investigate Plank's thermal radiation distribution function in more detail. Thermal radiation distributions for our Sun at 5,800 K and for a bonfire at 873 K and for a human body at 310 K are shown in Figures 1-1 and 1-2.

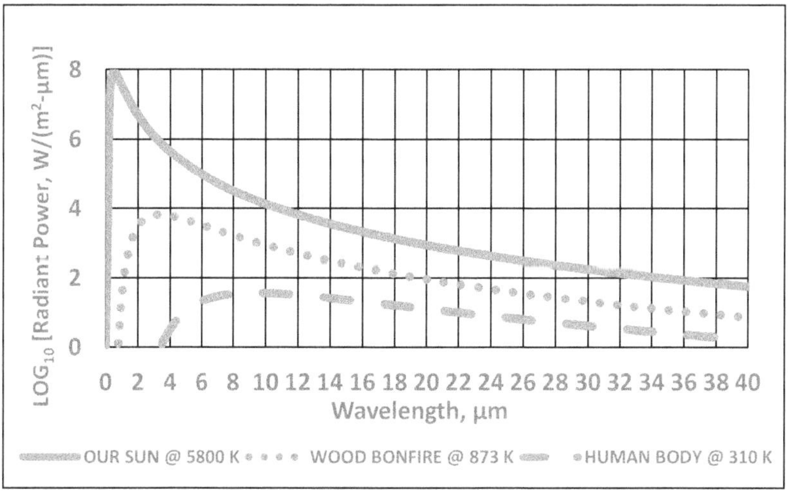

Figure 1-1 Radiant power of blackbodies at different surface temperatures versus wavelength

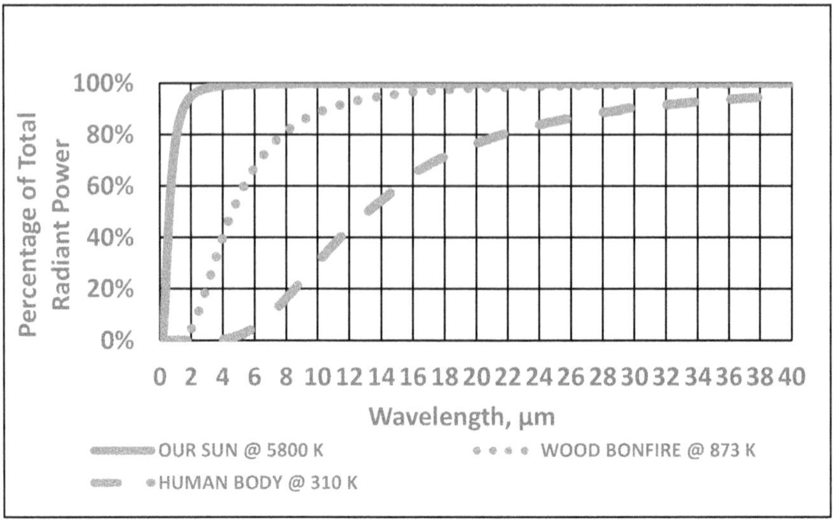

Figure 1-2 Percentage of total radiant power emitted by a blackbody versus wavelength for different surface temperatures

Maximum radiant power of our Sun occurs at λ=0.5 µm. Maximum radiant power of a wood bonfire occurs at λ=3.32 µm and for a human body it occurs at λ=9.35 µm.

90% of our Sun's radiant power is emitted at wavelengths less than 1.6 µm. 90% of a wood bonfire's radiant power is emitted at wavelengths less than 10.8 µm. 90% of our body's radiant power is emitted at wavelengths less than 30.3 µm.

Planck's distribution function for thermal radiation, given in Equation 1-7, can be rewritten as a function of a single variable, $X = \lambda \times T$, as shown in Equation 1-9. Equation 1-9 can be used conveniently to calculate radiant power emissions in wavelength bands.

$$\frac{E_b(0 \rightarrow X)}{\sigma \times T^4} = \int_0^X \left\{ \left[\frac{C_1}{\sigma \times X^5} \right] / \left[exp\left(\frac{C_2}{X} \right) - 1 \right] \right\} dX \qquad 1\text{-}9$$

Percentage of total thermal radiant power emitted by a blackbody versus $X = \lambda \times T$ is shown in Figure 1-3. 98.5% of the total thermal radiant power emitted by a blackbody occurs 0 < X < 20,000 µm-K.

The integrand in Equation 1-9 is plotted as a function of X in Figure 1-4. The integrand increases fast, namely high sensitivity to changes in X, from X = 0 to a maximum at X = 2,898 µm-K. Then it decreases slowly back to zero, namely low sensitivity to changes in X.

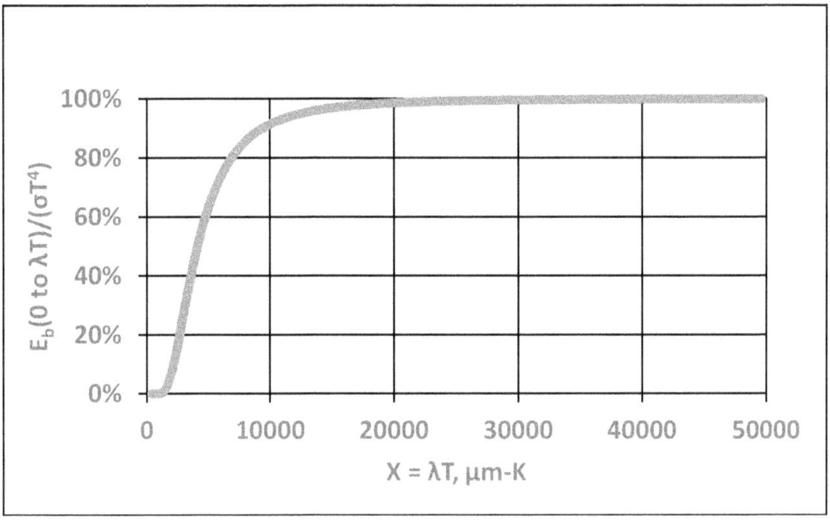

Figure 1-3 Percentage of total thermal radiant power emitted by a blackbody at temperature T, K in the wavelength interval 0 to X, μm

Now let us investigate sensitivities of Planck's thermal radiation distribution function, Equation 1-7, to wavelength at constant temperatures and also to temperature at constant wavelengths.

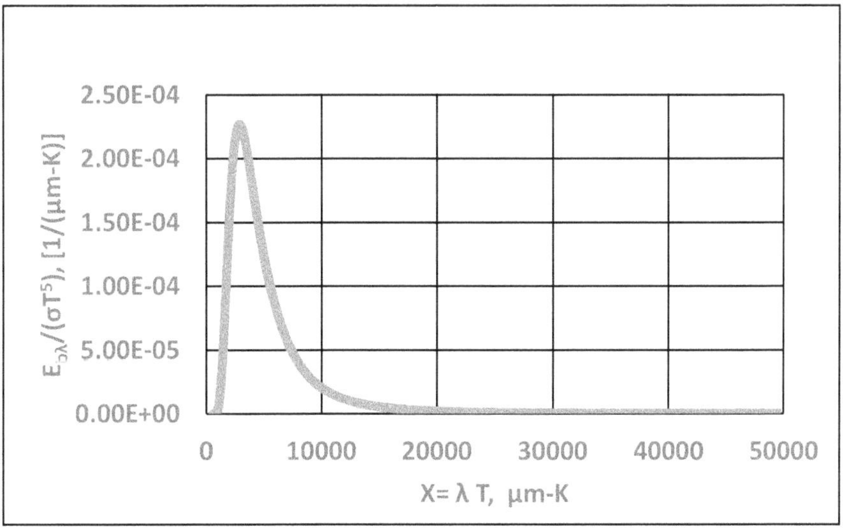

Figure 1-4 Integrand function in Equation 1-9 versus independent variable $X = \lambda \times T$

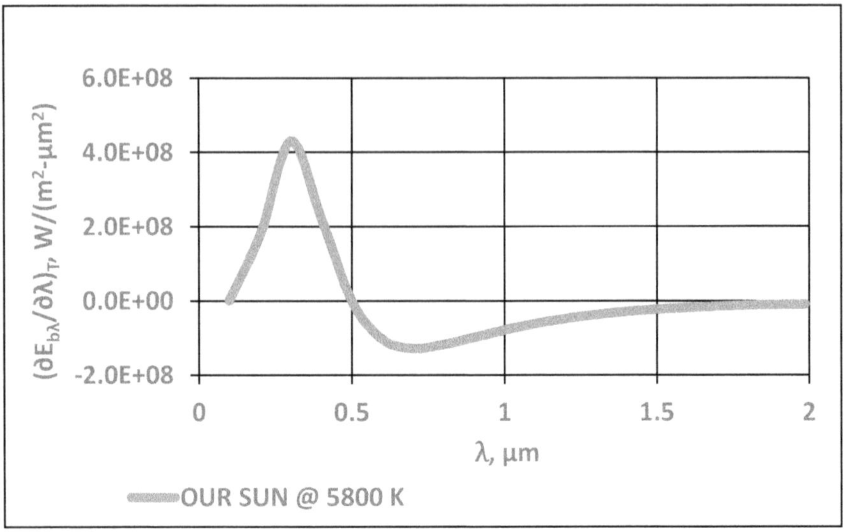

Figure 1-5 Sensitivity of Planck's thermal radiation distribution function to wavelength at Sun's surface temperature of 5800 K

Figure 1-5 shows sensitivities of thermal radiation emitted by a blackbody to wavelength at our Sun's surface temperature of 5,800 K. Sensitivities at this temperature range, around 5,800 K, are very large, namely at the order of 10^8 W/(m^2-μm^2). These sensitivities are positive below $\lambda = 0.5$ *μm*, they go through zero at $\lambda = 0.5$ *μm* and then they become negative. At these high temperatures, sensitivities of thermal radiation emitted by a blackbody to wavelength becomes negligible above $\lambda = 2$ *μm*.

Figure 1-6 shows sensitivities of thermal radiation emitted by a blackbody to wavelength at a wood bonfire's surface temperature of 873 K. Sensitivities at this temperature range, around 873 K, are moderate, namely at the order of 10^3 W/(m^2-μm^2). These sensitivities are positive below $\lambda = 3.32$ *μm*, they go through zero at $\lambda = 3.32$ *μm* and then they become negative. At these moderate temperatures, sensitivities of thermal radiation emitted by a blackbody to wavelength becomes negligible above $\lambda = 12$ *μm*.

Figure 1-7 shows sensitivities of thermal radiation emitted by a blackbody to wavelength at a human body's surface temperature of 310 K. Sensitivities at this temperature range, around 310 K, are very low, namely at the order of 1 W/(m^2-μm^2). These sensitivities are positive below $\lambda = 9.35$ *μm*, they go through zero at $\lambda= 9.35$ *μm* and then they become negative. At these low temperatures, sensitivities of thermal radiation emitted by a blackbody to wavelength becomes negligible above $\lambda = 32$ *μm*.

Figure 1-6 Sensitivity of Planck's thermal radiation distribution function to wavelength at wood bonfire's surface temperature of 873 K

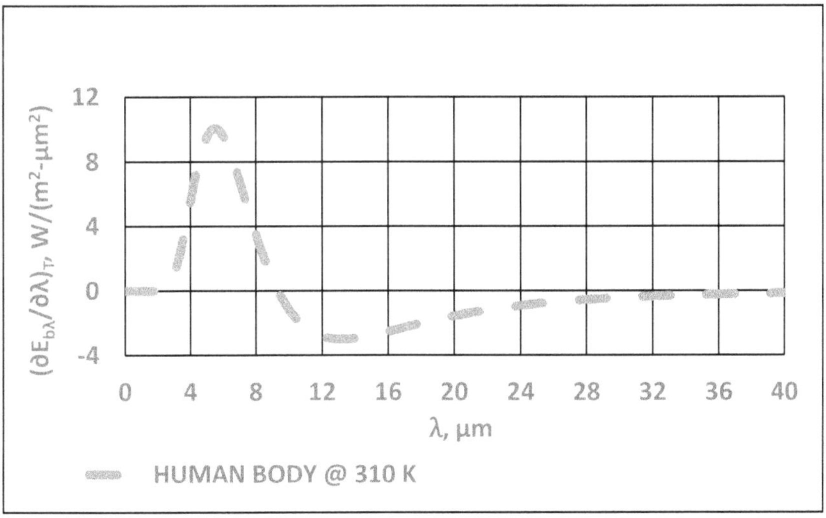

Figure 1-7 Sensitivity of Planck's thermal radiation distribution function to wavelength at a human body's surface temperature of 310 K

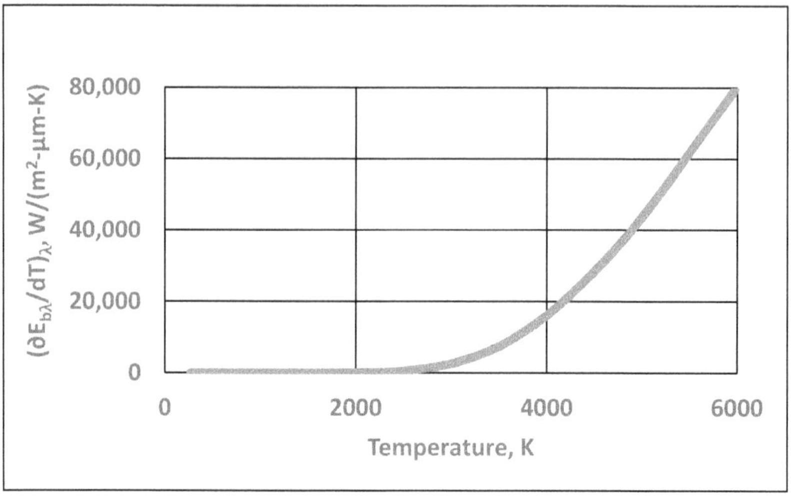

Figure 1-8 Sensitivity of Planck's thermal radiation distribution function to temperature at $\lambda = 0.5$ μm where Sun's maximum emissive power occurs

Figure 1-8 shows sensitivities of thermal radiation, $E_{b\lambda}$, emitted by a blackbody to temperature at the wavelength of 0.5 μm. Sensitivities at this wavelength increases as we go to higher temperatures, i.e. 6,000 K. These sensitivities at this wavelength are at the order of 10^3 W/(m^2-μm-K) These sensitivities go to zero below 2,000 K.

Figure 1-9 shows sensitivities of thermal radiation, $E_{b\lambda}$, emitted by a blackbody to temperature at the wavelength of 3.32 μm. Sensitivities at this wavelength

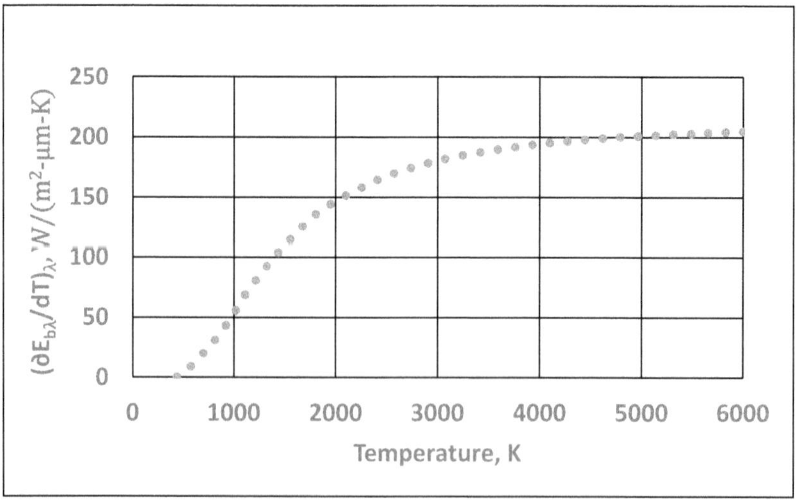

Figure 1-9 Sensitivity of Planck's thermal radiation distribution function to temperature at $\lambda = 3.32$ μm where wood bonfire's maximum emissive power occurs

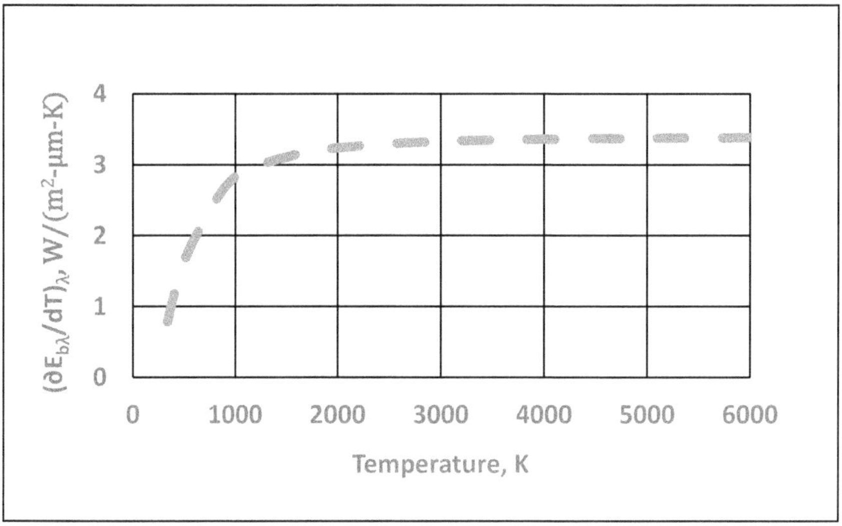

Figure 1-10 Sensitivity of Planck's thermal radiation distribution function to temperature at $\lambda = 9.35$ μm where a human body's maximum emissive power occurs

increases as we go to higher temperatures, i.e. 4,000 K. These sensitivities at this wavelength are at the order of 10^2 W/(m^2- μm-K) These sensitivities flatten out to 200 W/(m^2-μm-K) above 4,000 K.

Figure 1-10 shows sensitivities of thermal radiation, $E_{b\lambda}$, emitted by a blackbody to temperature at the wavelength of 9.35 μm. Sensitivities at this wavelength increases as we go to higher temperatures, i.e. 1,000 K. These sensitivities at this wavelength are at the order of 1 W/(m^2-μm-K) These sensitivities flatten out to 3.3 W/(m^2-μm-K) above 2,000 K.

We can also investigate Planck's distribution function for thermal radiation emitted by a blackbody at a certain frequency, f in Hz, and at a certain temperature T. Let us specify E_{bf} in W/(m^2-Hz) as the distribution function of thermal radiation emitted by a blackbody in the frequency space and at a certain temperature. The equivalence of thermal radiation emitted at an increment of wavelength dλ is equal to the thermal radiation emitted at an increment of frequency df as shown in Equation 1-10.

$$E_{b\lambda} \times d\lambda = - E_{bf} \times df \qquad 1\text{-}10$$

For electromagnetic waves moving in vacuum, their frequencies and wavelengths are related, namely $f = C/\lambda$. where $C = 3x10^8$ *m/s* is the speed of light in vacuum. E_{bf} is obtained from Equations 1-7 and 1-10 and the relationship between frequency and wavelength. The result it is shown in Equation 1-11.

$$E_{bf} = \left(\frac{c_1 \times f^3}{c^4}\right) / \left[exp\left(\frac{C_2 \times f}{c \times T}\right) - 1\right] \qquad 1\text{-}11$$

Thermal radiation emitted by a blackbody in the frequency space and at certain temperatures, i.e. Equation 1-11, is shown in Figure 1-11 for our Sun at 5,800 K, for a wood bonfire at 873 K and for a human body at 310 K. Thermal radiation emitted by our Sun's surface goes through a maximum at a frequency of $3.41\text{x}10^{14}$ Hz. Thermal radiation emitted by a wood bonfire's surface goes through a maximum at a frequency of $5.14\text{x}10^{13}$ Hz. Thermal radiation emitted by a human body's surface goes through a maximum at a frequency of $1.82\text{x}10^{13}$ Hz.

Percentage of total radiant power emitted by a blackbody surface versus frequency for three different surface temperatures are shown in Figure 1-12. 90% of our Sun's radiant power is emitted at frequencies greater than $2.25\text{x}10^{14}$ Hz. 90% of a wood bonfire's radiant power is emitted at frequencies greater than 2.82×10^{13} Hz. 90% of our body's radiant power is emitted at frequencies greater than 1.02×10^{13} Hz.

Sensitivity of E_{bf} to f at three different blackbody surface temperatures are shown in Figures 1-13, 1-14 and 1-15. These sensitivities show similar behavior, to $\left(\frac{\partial E_{b\lambda}}{\partial \lambda}\right)_T$ that are shown in Figures 1-5, 1-6 and 1-7 above. These sensitivities are zero at low frequencies and as frequency increases they increase positively.

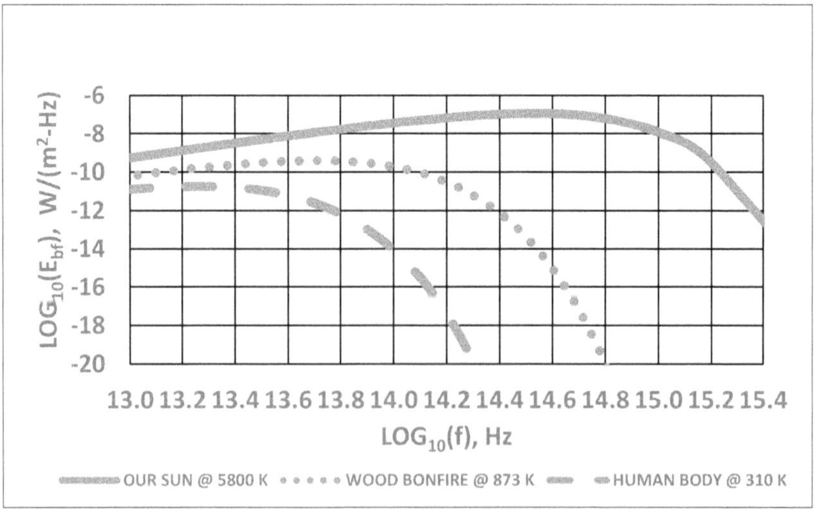

Figure 1-11 Thermal radiation emitted by a blackbody in the frequency space and at certain surface temperatures

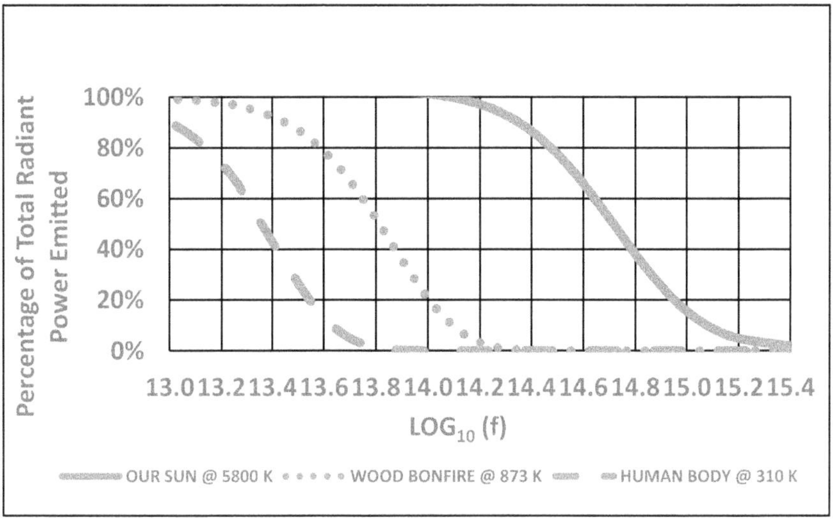

Figure 1-12 Percentage of total radiant power emitted by a blackbody versus frequency for different surface temperatures

Then they go through a maximum and then go to down to zero where the thermal radiation emitted by a blackbody in the frequency space and at a certain surface temperature is a maximum. Then these sensitivities become negative and go through a minimum as frequency increases and then they approach zero as frequency increases further.

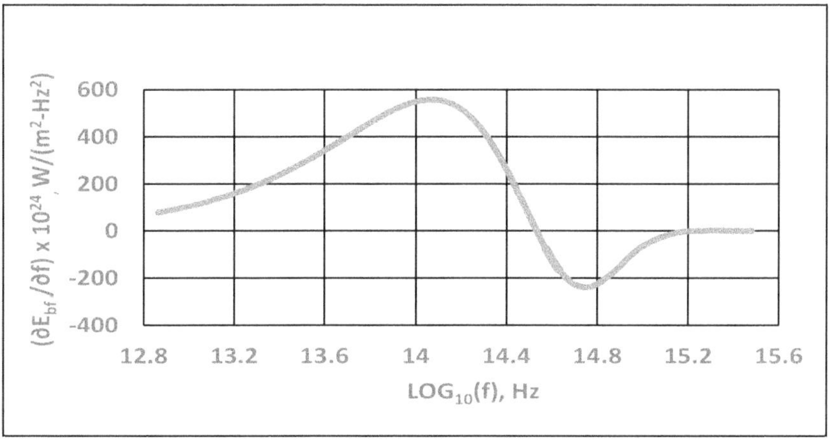

Figure 1-13 $(\partial E_{bf}/\partial f)_T \times 10^{24}$ versus $LOG_{10}(f)$ for our Sun's surface temperature at 5,800 K

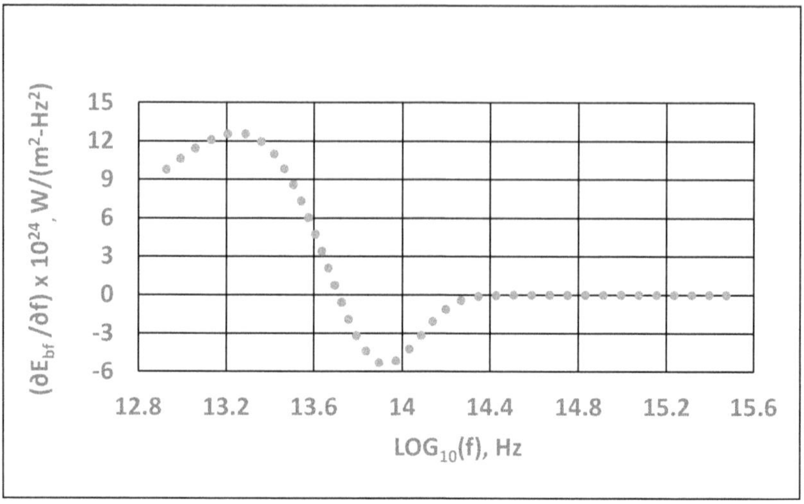

Figure 1-14 $(\partial E_{bf}/\partial f)_T \times 10^{24}$ versus $LOG_{10}(f)$ for wood bonfire's surface temperature at 873 K

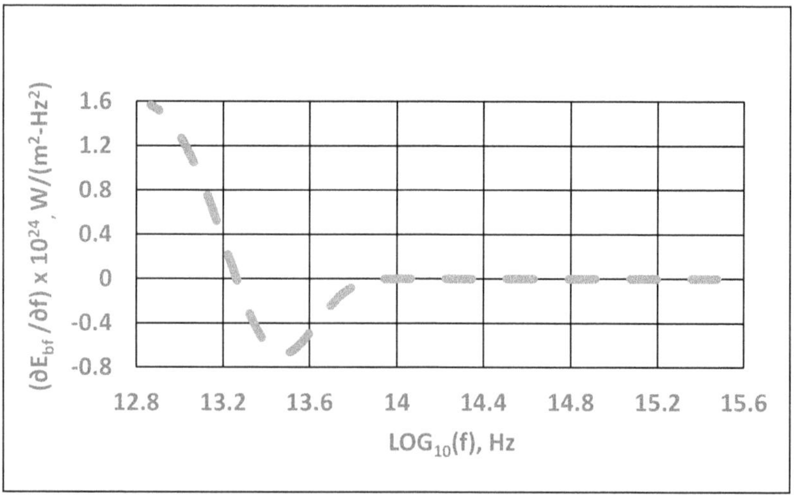

Figure 1-15 $(\partial E_{bf}/\partial f)_T \times 10^{24}$ versus $LOG_{10}(f)$ for a human body's surface temperature at 310 K

Figure 1-16 shows sensitivities of thermal radiation, E_{bf}, emitted by a blackbody to temperature at the frequency of 3.41×10^{14} Hz. This frequency is at the maximum E_{bf} at $T = 5{,}800$ K. Sensitivities at this frequency increases as we go to higher temperatures. These sensitivities at this frequency are at the order of 10^{-11} W/(m^2-Hz-K). These sensitivities go to zero below 2,000 K.

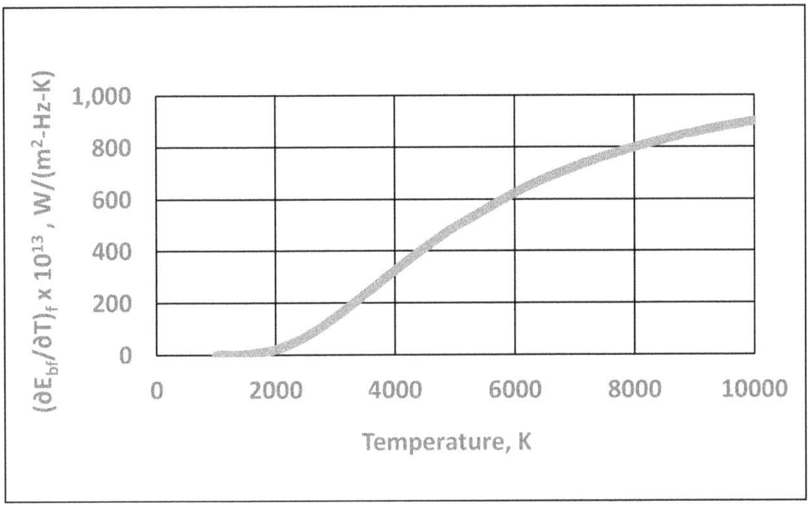

Figure 1-16 $(\partial E_{bf}/\partial T)_f \times 10^{13}$ versus temperature in K at f = 3.41×10^{14} Hz

Figure 1-17 shows sensitivities of thermal radiation, E_{bf}, emitted by a blackbody to temperature at two different frequencies of $5.14\text{x}10^{13}$ Hz and $1.82\text{x}10^{13}$ Hz. One frequency is at the maximum E_{bf} at T = 873 *K* and the other is at the maximum E_{bf} at T = 310 K. Sensitivities at these frequencies increase as we go to higher temperatures and then become a constant. Sensitivities at $5.14\text{x}10^{13}$ Hz is at the order of 10^{-12} W/(m^2-Hz-K). Sensitivities at 1.82×10^{13} Hz is at the order of 10^{-13} W/(m^2-Hz-K).

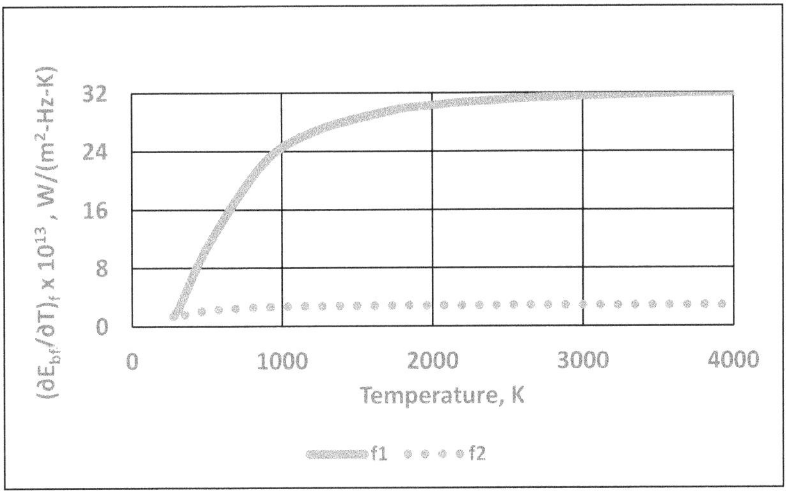

Figure 1-17 $(\partial E_{bf}/\partial T)_f \times 10^{13}$ versus temperature in K at f1 = 5.14×10^{13} Hz and f2 = 1.82×10^{13} Hz

2

Total Hemispherical Emissivity

In this chapter, we will deal with transient cooling of vertically hanged hot special steel plates by radiation and natural convection heat transfer. In engineering calculations of radiation heat transfer, a real surface emitting thermal radiation at a temperature T_s has a total emissive power less than the total emissive power of a black surface at T_s, namely less than $\sigma \times T_s^4$. The ratio of a real surface's emissive power to a black surface's emissive power at the same T_s is called the emissivity of a surface. Emissivity of a surface is a function of surface temperature, wavelength and direction of radiation emission. In engineering calculation, we deal with total hemispherical emissivity, ε_{Ts}, which is the emissivity of a surface that is averaged over all wavelengths and hemispherical directions and which is only a function of surface temperature as given in Equation 2-1.

$$\varepsilon_{Ts} = \frac{E(T_s)}{E_b(T_s)} \qquad \text{2-1}$$

where $E(T_S)$ is the real emissive power from a surface in Watts and $E_b(T_S) = \sigma \times T_s^4$ is the black surface's emissive power in Watts. For steel surfaces total hemispherical emissivity varies a lot due to its surface finish and surface oxidation. For the special steel that is considered in the present example, the total hemispherical emissivity is shown as a function of its surface temperature in Figure 2-1. The total hemispherical emissivity decreases from the initial high surface temperature at 1773 K to the final low surface temperature at 373 K. The average total hemispherical emissivity between 373 K and 1773 K is 0.17.

Let us investigate the transient cooling of vertically hanged hot special steel plates of different lengths by radiation and natural convection heat transfer using the temperature varying total hemispherical emissivity as shown in Figure 2-1. Let us also investigate the same calculations using an average total hemispherical emissivity of 0.17. Then we can determine errors in cooling times of special steel plates with different lengths when an average total hemispherical emissivity of 0.17 is used. Special steel plates have a thickness of 0.004 m and a width of 0.5 m.

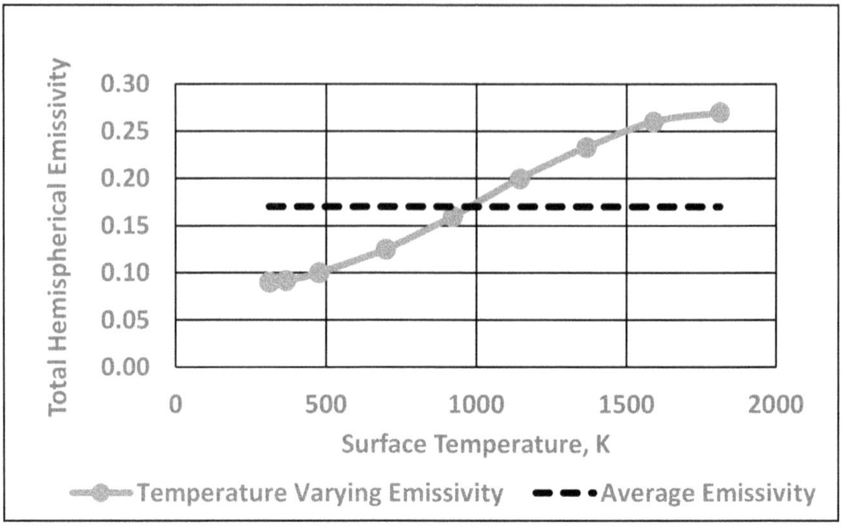

Figure 2-1 Total hemispherical emissivity, ε_{Ts}, versus surface temperature for special steel plates

Special steel plates come in different lengths starting from a low of 0.1 m to a high of 2.5 m.

The transient energy balance between the surfaces of a hanging special steel plate and the environment is given in Equation 2-2.

$$\left(\rho \times c_p \times V\right)_s \times \frac{dT_s}{d\theta} = -\varepsilon_{Ts} \times \sigma \times A_s \times \left(T_s^4 - T_{env}^4\right) - h_{nc} \times A_s \times \left(T_s - T_{env}\right) \quad \text{2-2}$$

where $(\rho \times c_p \times V)_s$ is the thermal capacitance of the special steel plate in (W-s)/K. T_s is the surface temperature of the special steel plate which is initially at 1773 K and cools down to a final temperature of 373 K. θ is the time in seconds. σ is the Stefan-Boltzmann constant at 5.67×10^{-8} W/(m^2-K^4). A_s is the total surface area of the special steel plate in m^2. T_{env} is the constant environment and air temperatures at 20 C. h_{nc} is the natural convection heat transfer coefficient in W/(m^2-K) for vertical plate cooling and it is given in Equation 2-3, (see Ref. 13).

$$\overline{Nu_L} = \frac{h_{nc} \times L}{k_{air}} = \left\{0.825 + \frac{0.387 \times Ra_L^{1/6}}{\left[1 + \left(0.492 / Pr_{air}\right)^{9/16}\right]^{8/27}}\right\}^2 \quad \text{2-3}$$

where L is the length of the special steel plate in m. k_{air} is the thermal conductivity of surrounding air in W/(m-K). Pr_{air} is the Prandtl number of air, namely the ratio of air's momentum diffusivity to its thermal diffusivity. Ra_L is the Rayleigh number and it is given in Equation 2-4.

$$Ra_L = \frac{g \times \beta_{air} \times (T_s - T_{env}) \times L^3}{v_{air} \times \alpha_{air}} \qquad 2\text{-}4$$

where all air properties are taken at the film temperature of $T_{film} = 0.5 \times (T_s + T_{env})$. g is the Earth's gravitational acceleration at sea level, namely 9.81 m/s^2. $\beta_{air} = \frac{1}{T_{film}}$ is the volumetric thermal expansion coefficient in 1/K. v_{air} is air's momentum diffusivity in m^2/s. α_{air} is air's thermal diffusivity in m^2/s.

Typical air properties that are used in the present calculations at the film temperature of 900 K and 1 atmosphere are given below:

$\beta_{air} = 0.0011\ 1/K$, $k_{air} = 0.062\ W/(m-K)$, $Pr_{air} = 0.72$,

$v_{air} = 1.03 \times 10^{-4}\ m^2/s$, $\alpha_{air} = 1.43 \times 10^{-4}\ m^2/s$

Equation 2-2, a transient non-linear differential equation, is solved using an explicit finite difference method with a time interval of 10 seconds and an initial condition of $T_s = 1773\ K$ as shown in Equation 2-5.

$$T_{s-(i+1)} = T_{s-(i)} - \left[\frac{\theta_{i+1} - \theta_i}{(\rho \times c_p \times V)_s}\right] \times \left[\varepsilon_{Ts}(T_i) \times \sigma \times A_s \times (T_{s-(i)}^4 - T_{env}^4) + h_{nc} \times A_s \times (T_{s-(i)} - T_{env})\right] \qquad 2\text{-}5$$

Using Equations 2-1 through 2-5 and temperature varying total hemispherical emissivity, ε_{Ts}, given in Figure 2-1, cooling temperatures for a 1 meter long special steel plate are calculated using 1 second and 10 second time intervals as shown in Figure 2-2. Cooling temperature calculations using the 10-second time interval is 3.2% lower than the 1-second time interval calculations at high temperatures around 1500 K. This temperature difference decreases fast and goes down to 1.7% around 800 K. Therefore, in all the following analyses, a 10-second time interval is used. A 2.5 m long special steel plate cools down from 1773 K to 373 K in 2690 seconds (44 minutes and 50 seconds).

Special steel plate's surface temperature, with a length of 2 meters, width of 0.5 meter and thickness of 0.004 meter, cools down from 1773 K to 373 K in 2660 seconds (44 minutes and 20 seconds) as shown in Figure 2-3. Total hemispherical emissivity that is used in these calculations varies with surface temperature as shown in Figure 2-1. Up to 340 seconds (5 minutes and 40 seconds) of cooling time, heat loss due to radiation dominates the cooling event. After 340 seconds natural convection heat loss is higher than the radiation heat loss as depicted in Figure 2-4.

Special steel plate's surface temperature, with a length of 2 meters, width of 0.5 meter and thickness of 0.004 meter, cools down from 1773 K to 373 K in 2326 seconds (38 minutes and 46 seconds) as shown in Figure 2-5. Total hemispherical

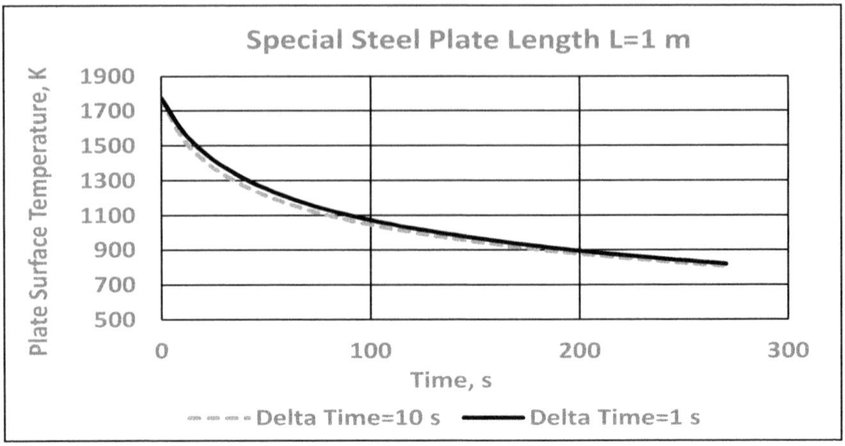

Figure 2-2 Plate surface temperatures for a 1 meter long special steel plate using 1-second time intervals and 10-second time intervals using surface temperature dependent total hemispherical emissivity given in Figure 2-1

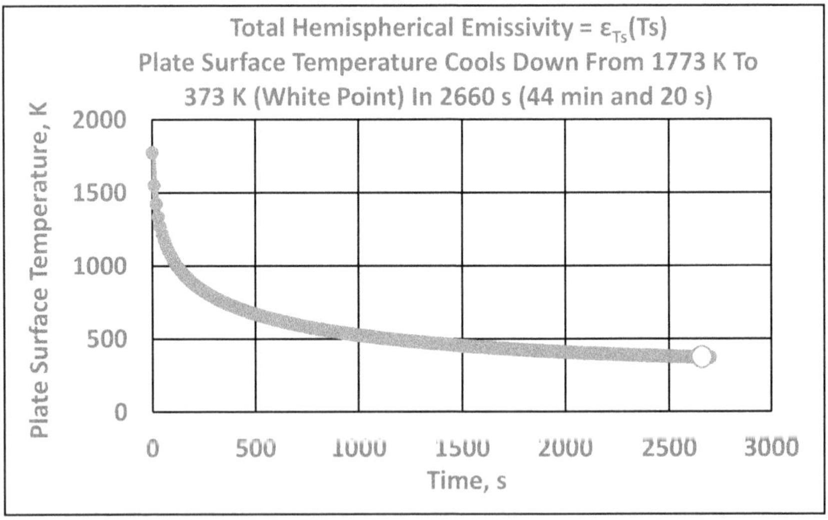

Figure 2-3 Special steel plate's surface temperature, with a length of 2 meters, width of 0.5 meter and thickness of 0.004 meter, cools down from 1773 K to 373 K in 2660 seconds (44 minutes and 20 seconds) using surface temperature dependent total hemispherical emissivity given in Figure 2-1

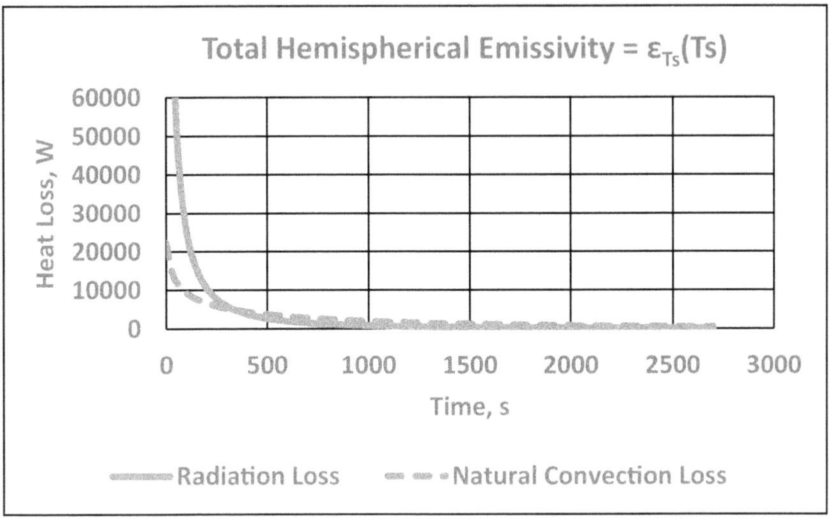

Figure 2-4 Heat loss from the special plate's surfaces due to radiation and natural convection heat transfers versus cooling time for a plate with a length of 2 meters, width of 0.5 meter and thickness of 0.004 meter using surface temperature dependent total hemispherical emissivity given in Figure 2-1. Up to 340 seconds (5 minutes and 40 seconds) of cooling time, heat loss due to radiation dominates the cooling event.

emissivity that is used in these cooling calculations has an average value of 0.17 and it does not vary with surface temperature as shown in Figure 2-1. Up to 500 seconds (8 minutes and 20 seconds) of cooling time, heat loss due to radiation dominates the cooling event. After 500 seconds natural convection heat loss is higher than the radiation heat loss as depicted in Figure 2-6.

For a 2 meters long special steel plate, the cooling time is 12.6% lower when an average total hemispherical emissivity is used during the cooling calculations instead of temperature varying ones as shown in Figure 2-1. For engineering calculations, it is recommended to use surface temperature dependent total hemispherical emissivities for more accurate results. Cooling times for different length special steel plates are shown in Figure 2-7 using temperature dependent total hemispherical emissivities and also using an average total hemispherical emissivity of 0.17 as given in Figure 2 1. Using an average total hemispherical emissivity of 0.17 creates cooling time errors varying from 8.6% for the small length plates up to 12.7% for the long length plates.

Sensitivities of cooling times to plate lengths are shown in Figure 2-8 for temperature dependent total hemispherical emissivities and for an average total hemispherical emissivity of 0.17 as given in Figure 2-1. Both sensitivity curves behave similarly. Sensitivities of cooling times to plate lengths are high at small

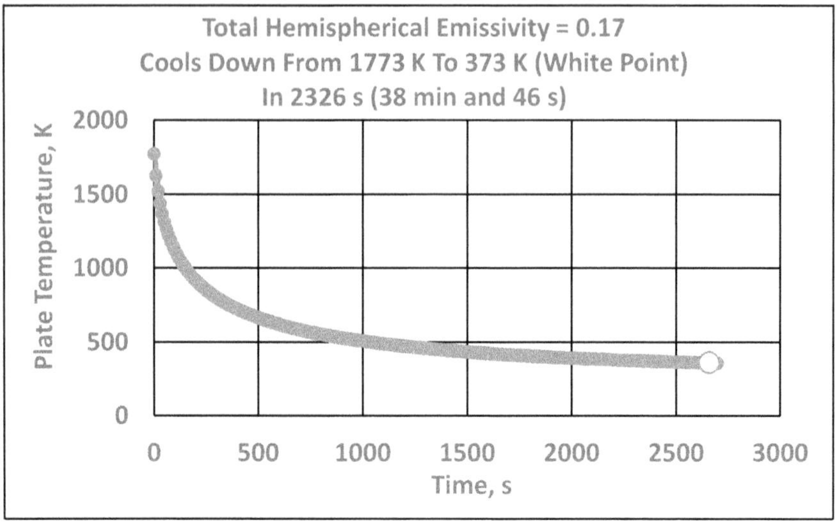

Figure 2-5 Special steel plate's surface temperature, with a length of 2 meters, width of 0.5 meter and thickness of 0.004 meter, cools down from 1773 K to 373 K in 2326 seconds (38 minutes and 46 seconds) using an average total hemispherical emissivity of 0.17 given in Figure 2-1

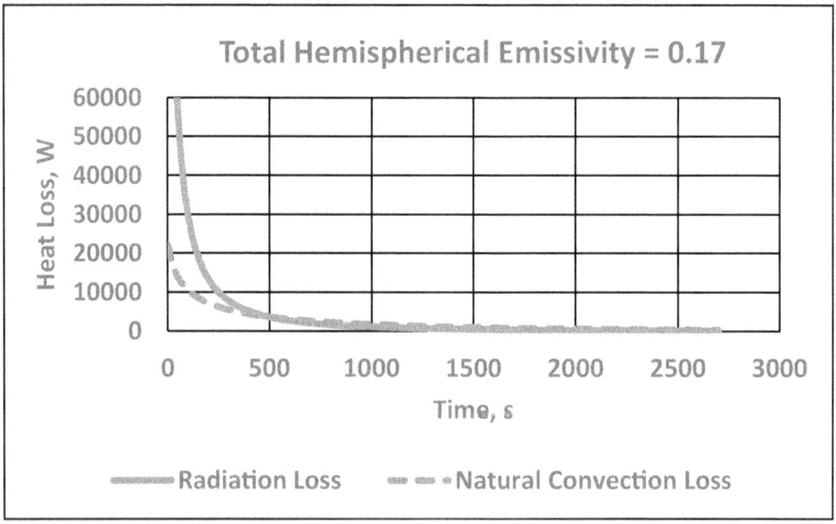

Figure 2-6 Heat loss from the special plate's surfaces due to radiation and natural convection heat transfers versus cooling time for a plate with a length of 2 meters, width of 0.5 meter and thickness of 0.004 meter using an average total hemispherical emissivity of 0.17 given in Figure 2-1. Up to 500 seconds (8 minutes and 20 seconds) of cooling time, heat loss due to radiation dominates the cooling event.

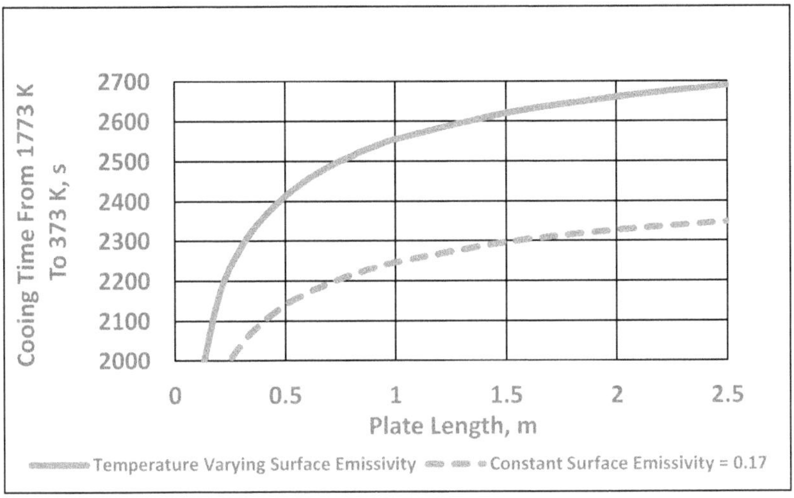

Figure 2-7 Cooling times for different length special steel plates using temperature dependent total hemispherical emissivities and using an average total hemispherical emissivity of 0.17 as given in Figure 2-1.

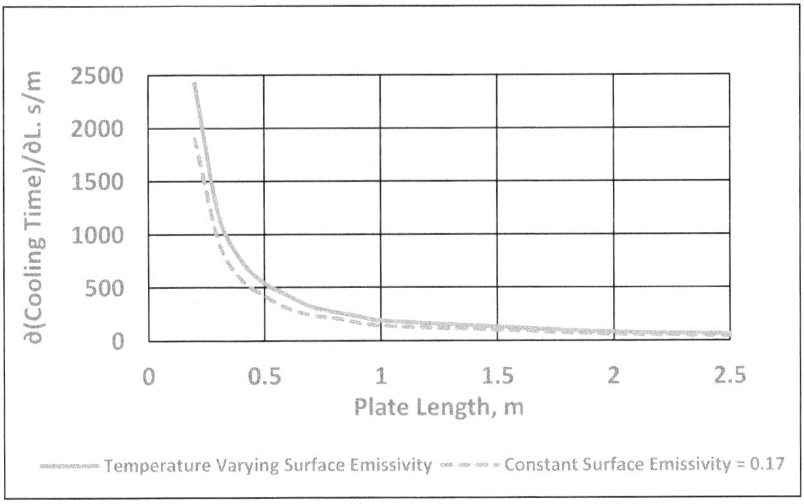

Figure 2-8 Sensitivities of cooling times to plate lengths for temperature dependent total hemispherical emissivities and for an average total hemispherical emissivity of 0.17 as given in Figure 2-1.

plate lengths, namely at shorter cooling times. Sensitivities of cooling times to plate lengths approach zero for both cases at L = 2.5 m.

3

Thermal Radiation Configuration Factor Example

In this chapter, we will deal with thermal radiation exchange between two black-bodies which are perfect electromagnetic radiation absorbers and emitters. In order to determine the net exchange of thermal radiation between two black surfaces, we only have to know the configuration factor, $F_{A1\text{—}A2}$, between these two surfaces and their surface temperatures, T_{A1} and T_{A2}, as shown in Equation 3-1. $A1$ is the surface area of body number one. σ is the Stefan-Boltzmann constant which is 5.67×10^{-8} W/m^2-K^4. The configuration factor is defined as the fraction of thermal radiation that leaves a surface and intercepted by another surface. Configuration factors are sometimes called view factors or shape factors in literature, see Refs [13] and [19]. Derivation of configuration factors which is a geometric relationship between surfaces can be found extensively in literature, Ref [33].

$$Q_{A1-A2} = A1xF_{A1-A2}x\sigma x\left(T_{A1}^4 - T_{A2}^4\right) \qquad 3\text{-}1$$

Configuration factor, $F_{dA1\text{-}dA2}$, between a surface area element dA1 and a surface area element dA2 is given in Equation 3-2 and it is defined by the length of the line, S, that connects these two elements, by cosines of angles between normal directions to these surface area elements and the line S, namely cos(ϕ_1) and cos(ϕ_2), and by areas, dA1 and dA2, of these surface elements, see Ref [33].

$$F_{dA1-dA2} = \frac{cos(\varphi_1)xcos(\varphi_2)xdA2}{\pi xS^2} \qquad 3\text{-}2$$

When finite surface areas are considered, integration of Equation 3-2 in order to get a closed form algebraic equation can be very complicated. Then numerical integration of Equation 3-2 can be used with high accuracy results. If we want to obtain the configuration factor between a surface element dA1 and a finite surface area A2, we have to integrate Equation 3-2 with respect to dA2 as shown in Equation 3-3.

$$F_{dA1-A2} = \int_{A2}\left[\frac{cos(\varphi_1)xcos(\varphi_2)xdA2}{\pi xS^2}\right] \qquad 3\text{-}3$$

If we want to obtain the configuration factor between a finite surface A1 and a finite surface area A2, we have to integrate Equation 3-3 with respect to dA1 as shown in Equation 3-4.

$$F_{A1-A2} = \frac{1}{A_1}\int_{A1}\left\{\int_{A2}\left[\frac{cos(\varphi_1)xcos(\varphi_2)xdA2}{\pi xS^2}\right]\right\}dA1 \qquad 3\text{-}4$$

Also, we can determine one configuration factor from the other by using the reciprocity expressions given in Equations 3-5a, 3-5b and 3-5c.

$$dA1\ x\ F_{dA1\_dA2} = dA2\ x\ F_{dA2\text{-}dA1} \qquad 3\text{-}5a$$

$$dA1\ x\ F_{dA1\text{-}A2} = A2\ x\ F_{A2\text{-}dA1} \qquad 3\text{-}5b$$

$$A1\ x\ F_{A1\text{-}A2} = A2\ x\ F_{A2\text{-}A1} \qquad 3\text{-}5c$$

Now let us calculate the configuration factor from a cylinder with a small diameter, dA1, that is perpendicular to the center of a disk, A2, as shown in cross-section view in Figure 3-1.

The configuration factor from an element on a small diameter cylinder, dA1, that is perpendicular to the center of a disk element, dA2, as shown in Figure 3-1 can be written as shown in Equation 3-6.

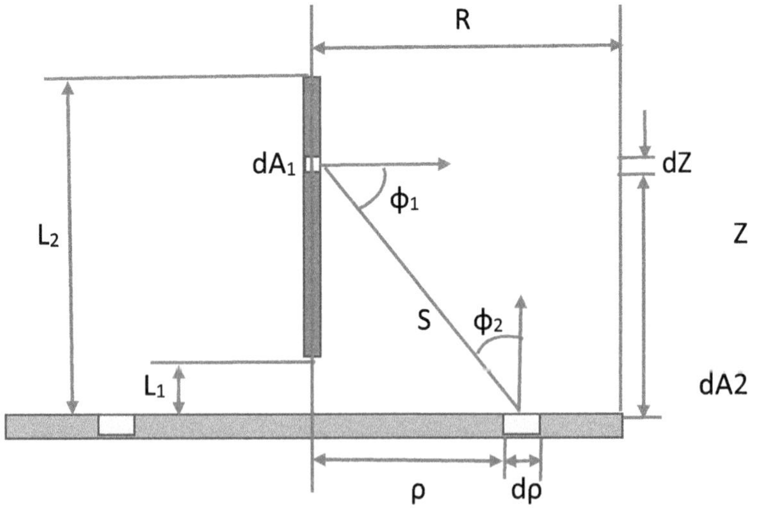

Figure 3-1 Cross-sectional geometric for the configuration factor from a cylinder with a small diameter, dA1, that is perpendicular to the center of a disk, A2

$$F_{dA1-dA2} = \frac{2 x \rho^2 x\ z\ x\ d\rho}{\pi x S^4} \qquad 3\text{-}6$$

where $S^2 = p^2 + Z^2$. After integrating Equation 3-6 from $\rho = 0$ to $\rho = R$, we obtain the following algebraic equation for $F_{dA1\text{-}A2}$ as shown in Equation 3-7.

$$F_{dA1-A2} = -\frac{RxZ}{\pi x\left(R^2 + Z^2\right)} + \left(\frac{1}{\pi}\right) x\ tan^{-1}\left(\frac{R}{Z}\right) \qquad 3\text{-}7$$

Next we can integrate Equation 3-7 as shown in Equation 3-4 between $Z = L_1$ and $Z = L_2$ in order to obtain the configuration factor between a cylinder with a small diameter and height $(L_2 - L_1)$, namely surface A1, that is perpendicular to the center of a disk with a radius R, namely surface A2. The result is shown in Equation 3-8.

$$F_{A1-A2} = \left(\frac{1}{\pi}\right) x \left[\left(\frac{L_2}{L_2 - L_1}\right) x\ tan^{-1}\left(\frac{R}{L_2}\right) - \left(\frac{L_1}{L_2 - L_1}\right) x\ tan^{-1}\left(\frac{R}{L_1}\right)\right] \qquad 3\text{-}8$$

If we let $(L_2 - L_1) = H$ where H is the height of the cylinder with a small diameter, then Equation 3-8 can be rewritten as follows in Equation 3-9.

$$F_{A1-A2} = \left(\frac{1}{\pi}\right) x \left[\left(\frac{L_1}{H} + 1\right) x\ tan^{-1}\left(\frac{R}{L_1 + H}\right) - \left(\frac{L_1}{H}\right) x\ tan^{-1}\left(\frac{R}{L_1}\right)\right] \qquad 3\text{-}9$$

For the special case where $L_1 = 0$, Equation 3-9 simplifies to Equation 3-10.

$$F_{A1-A2} = \left(\frac{1}{\pi}\right) x\ tan^{-1}\left(\frac{R}{H}\right) \qquad 3\text{-}10$$

For the special case where $L_1 = 0$, as $R \rightarrow 0$ or as $H \rightarrow \infty$, $F_{A1\text{-}A2} \rightarrow 0$. Also, for the special case where $L_1 = 0$, as $R \rightarrow \infty$ or as $H \rightarrow 0$, $F_{A1\text{-}A2} \rightarrow 0.5$. Plots of FA1-A2 versus R/H for results given in Equations 3-9 and 3-10 are shown in Figure 3-2. As the disk radius increases, FA1-A2 approaches the limiting value of 0.5. Also as L1/H increases, the cylinder's surface sees less of the disk surface, and therefore FA1-A2 approaches the limiting value of 0.5 slowly with increasing disk radius.

Sensitivities of FA1-A2 to disk radius, R/H, for different values of L1/H are shown in Figure 3-3. For L1/H = 0, the sensitivity of FA1-A2 to R/H starts at a value of 0.318 and decreases continuously towards zero as the disk radius increases. When the small diameter cylinder is separated from the disk surface, namely L1/H > 0, the sensitivities start at zero, increase to a maximum and then decrease towards zero. Also, sensitivities of FA1-A2 to R/H diminish as L1/H increases.

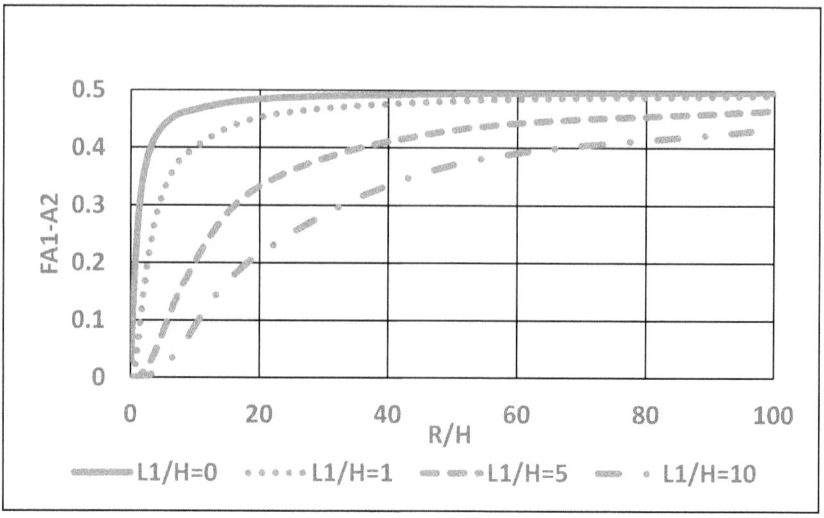

Figure 3-2 Configuration factors FA1-A2 versus R/H from a small diameter cylinder, A1, that is perpendicular to the center of a disk, A2, for different L1/H values

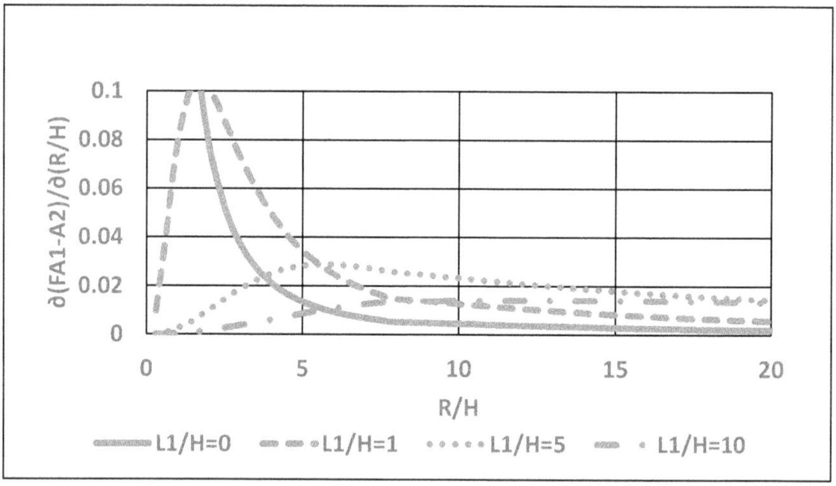

Figure 3-3 Sensitivity of FA1-A2 to R/H for different L1/H values

Plots of FA1-A2 versus L1/H for results given in Equations 3-9 and 3-10 are shown in Figure 3-4. As the cylinder distance to the disk, L1, increases, FA1-A2 approaches the limiting value of 0.0. Also as R/H increases, the cylinder's surface sees more of the disk's surface and therefore FA1-A2 approaches the limiting value of 0.0 slowly with increasing cylinder distance to the disk.

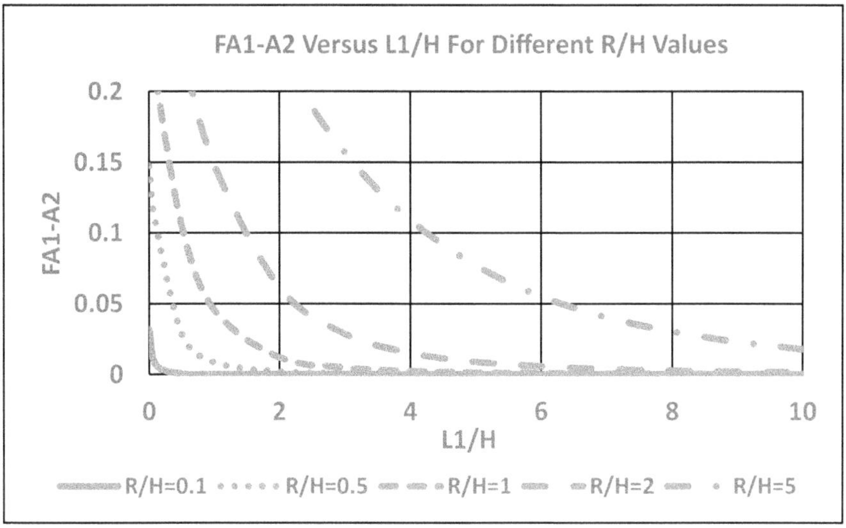

Figure 3-4 Configuration factors FA1-A2 versus L1/H from a small diameter cylinder, A1, that is perpendicular to the center of a disk, A2, for different R/H values

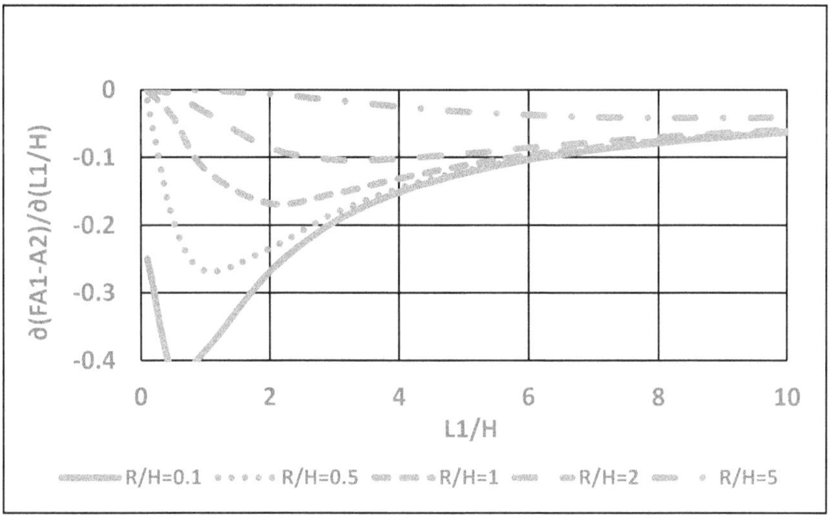

Figure 3-5 Sensitivity of FA1-A2 to L1/H for different R/H values

Sensitivity of FA1-A2 to cylinder distance to the disk surface, L1/H, for different values of R/H are shown in Figure 3-5. Sensitivities start at zero and its absolute value increases to a maximum and then its absolute value decreases towards zero as the cylinder distance to the disk surface increases. With increasing disk radius, sensitivity of FA1-A2 to cylinder distance to the disk surface, L1/H, decreases.

4

Vertical Radiant Heater

In this chapter, we will investigate the effectiveness of a rectangular and vertical thermal radiation heater to a rectangular and vertical elemental surface that is located at the base, Z=0, of the heater as shown in Figure 4-1. The elemental surface can be located at different X and Y locations at the base of the thermal radiation heater depending upon the desired temperature of the elemental surface. The thermal radiation heater A_2 has a rectangular surface with a height, H, and a width, W, where W<<H. The vertical elemental surface dA_1 has a rectangular surface with a height, H and a width, W.

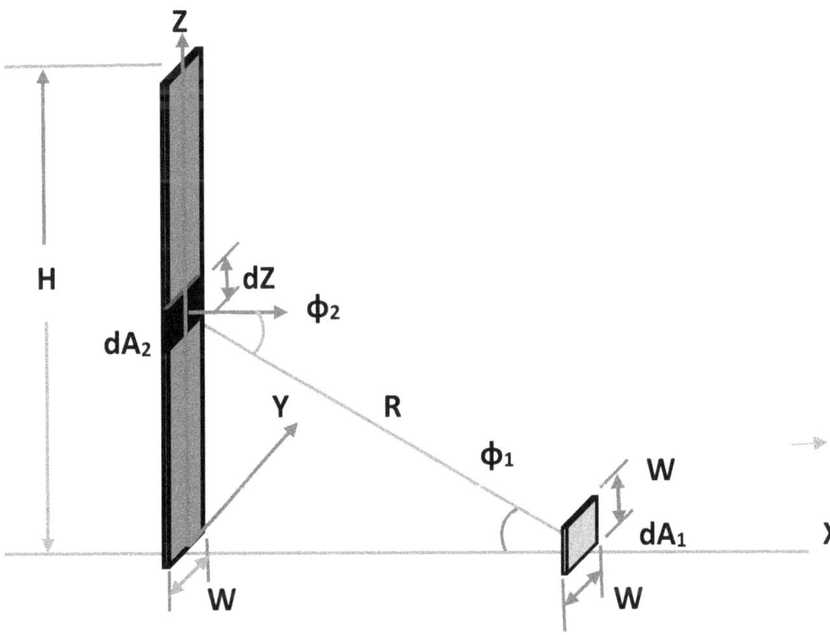

Figure 4-1 Rectangular and vertical thermal radiation heater, A_2 facing a rectangular and vertical elemental surface, dA_1

Assuming that both surfaces emit radiation diffusively, namely radiation intensity does not vary with emission angle, and also both surfaces are assumed to be black surfaces, namely all incident radiation on a surface is absorbed. The view factor between elemental area dA_1 and an elemental area dA_2 can be written as shown in Equation 4-1. See Ref [33] for view factor derivation.

$$F_{dA1\to dA2} = \frac{cos\varphi_1 \times cos\varphi_2}{\pi R^2} dA_2 \qquad \text{4-1}$$

where $R = \sqrt{X^2 + Y^2 + Z^2}$. However, from Figure 4-1,

$$cos\varphi_1 = cos\varphi_2 = \frac{X}{R}$$

Then Equation 4-1 takes the following form.

$$F_{dA1\to dA2} = \frac{X^2}{\pi R^4}(WdZ) \qquad \text{4-2}$$

Equation 4-2 can be integrated from Z=0 to Z=H to cover thermal radiation heater's view factor from its whole surface. Using the standard integral tables, Equation 4-2 is integrated to obtain the view factor between dA_1 and A_2 as shown in Equation 4-3.

$$F_{dA1\to A2} = \frac{WX^2}{2\pi}\left[\frac{H}{S^2\left(S^2 + H^2\right)} + \left(\frac{1}{S^3}\right) \times tan^{-1}\left(\frac{H}{S}\right)\right] \qquad \text{4-3}$$

where $S^2 = X^2 + Y^2$. Using the following reciprocity relationship between view factors shown in Equation 4-4, $F_{A2\to dA1}$ can be obtained as follows. See Ref [33] for view factor reciprocity relationships.

$$dA_1 \times F_{dA1\to A2} = A_2 \times F_{A2\to dA2} \qquad \text{4-4}$$

$F_{A2\to dA1}$ is given below in Equation 4-5.

$$F_{A2\to dA1} = \frac{c^2 M^2}{2\pi}\left[\frac{1}{P^2\left(P^2 + 1\right)} + \left(\frac{1}{P^3}\right) \times tan^{-1}\left(\frac{1}{P}\right)\right] \qquad \text{4-5}$$

where $C = \frac{W}{H}$, $M = \frac{X}{H}$, $N = \frac{Y}{H}$, and $P^2 = M^2 + N^2$. I would like to emphasize the assumption that $C \ll 1$ so that dA_1 can be treated as an elemental rectangular surface.

Let us analyze the view factor given in Equation 4-5 for a specific case where $H = 5$ *meters*, $W = 0.2$ *meter* or $W/H = 0.04$. Figure 4-2 shows view factors $F_{A2\to dA1}$ as a function of $M = \frac{X}{H}$ for different values of $N = \frac{Y}{H}$. View

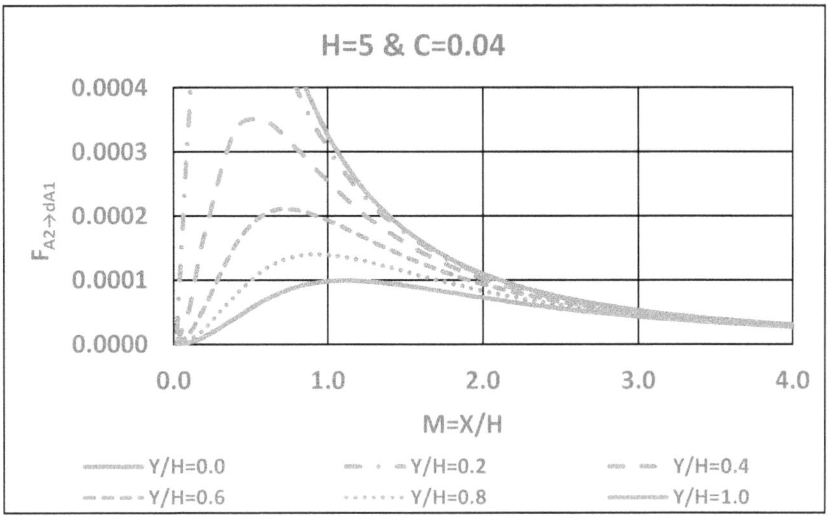

Figure 4-2 View factors $F_{A2 \to dA1}$ as a function of $M = \dfrac{X}{H}$ for different values of $N = \dfrac{Y}{H}$

factors from the thermal radiation heater to an elemental surface that is located at the base of the heater start at zero when $X = 0$ except for the $Y = 0$ case. Then as X increases view factors go through a maximum and then decrease back to zero as R increases. The maximum of the view factor decreases with increasing Y/H. For large X values, i.e. $X/H > 3$ or $X > 15$ *meters* for the present example as shown in Figure 4-2, effect of increasing Y/H diminishes. For the $Y = 0$ case, maximum of view factor $F_{A2 \to dA1}$ is 0.04 which corresponds to $F_{dA1 \to dA2} = 1.0$

Sensitivity of $F_{A2 \to dA1}$ to radiant heater height, H, at $Y = 0$ m is analyzed for different X values for the specific case given in Figure 4-3. For low H values, i.e. $H < 2$, sensitivity of $F_{A2 \to dA1}$ to radiant heater height, H, increases and goes through a maximum. Then sensitivity of $F_{A2 \to dA1}$ to radiant heater height decreases and asymptotes to a finite value. For high H values, i.e. $H > 2$, sensitivity of $F_{A2 \to dA1}$ to radiant heater height, H, increases continuously and asymptotes to a finite value without going through a maximum.

Let us now investigate the thermal radiation effect of the rectangular and vertical heater on the rectangular and vertical elemental surface, dA_1. It is assumed that there is no heat loss to the back of the elemental surface. The steady-state heat transfer balance for elemental surface, dA_1, is given in Equation 4-6 assuming that both surfaces emit radiation diffusively and they are black.

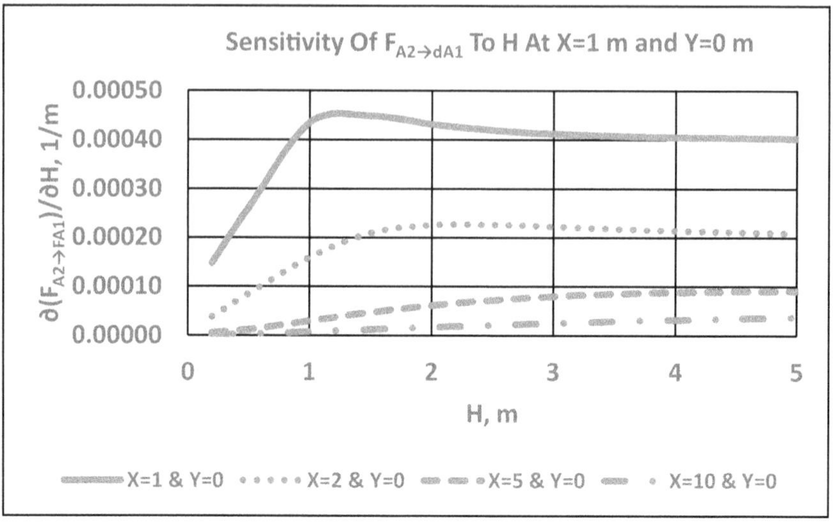

Figure 4-3 Sensitivity of $F_{A2\rightarrow dA1}$ to radiant heater height, H, at $Y = 0$ *meter* for different X values in meters

$$\sigma A_2 F_{A2\rightarrow dA1}\left(T_2^4 - T_1^4\right) = h_c dA_1\left(T_1 - T_{env}\right) + \sigma\, dA_1 F_{dA1\rightarrow env}\left(T_1^4 - T_{env}^4\right) \qquad 4\text{-}6$$

where $\sigma = 5.67x10^{-8}$ $W/(m^2 - K^4)$ is the Stefan Boltzmann constant (see Ref [32] for derivation of the Stefan-Boltzmann constant), h_c is the convection heat transfer coefficient between dA_1 and the environment in W/(m²-K), the view factor $F_{dA1\rightarrow env}$ is between the elemental surface and the environment, and T_{env} is the environment temperature in K. In Equation 4-6, left hand side of the equal sign represents the steady-state thermal radiation heat transfer from the radiation heater to the elemental surface. First term on the right-hand side of Equation 4-6 represents the steady-state convection heat transfer from the elemental surface to the environment. Second term on the right-hand side of Equation 4-6 represents the steady-state thermal radiation heat transfer from the elemental surface to the environment.

With the following parameters and using Equation 4-6, the maximum required view factor, $F_{A2\rightarrow dA1}$, to keep the dA_1 at a desired temperature level can be calculated.

$T_2 = 1{,}000\ C = 1273\ K,\ A_2 = 5\ m x\ 0.2\ m = 1\ m^2$

$h_c = 12\ W/(m^2 = K),\ dA_1 = 0.2\ m \times 0.2\ m = 0.04\ m^2$

$T_{env} = 20\ C = 293\ K,\ F_{dA1\rightarrow env} = 1.0$

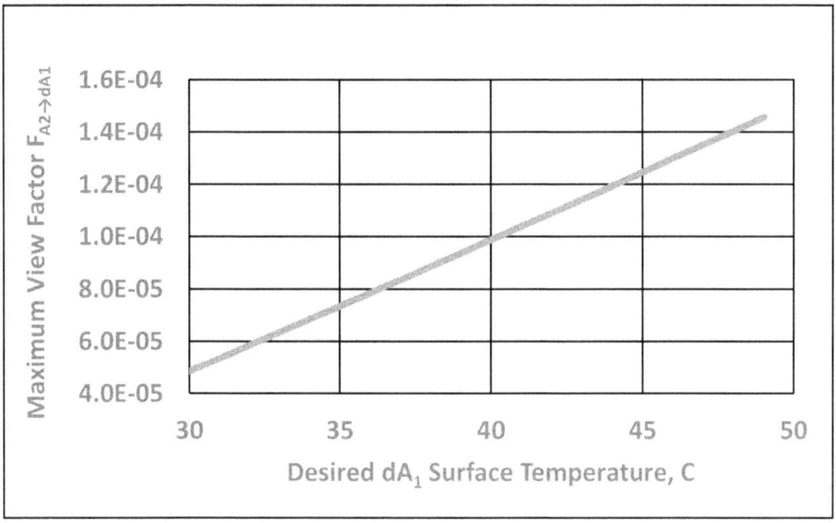

Figure 4-4 Maximum view factors, $F_{A2\rightarrow dA1}$, versus desired dA_1 surface temperature

Maximum required view factors, $F_{A2\rightarrow dA1}$, from the thermal radiation heater to the elemental surface, dA_1, temperatures are shown in Figure 4-4. As the dA_1 temperature increases, the view factor from the thermal radiation heater increases as expected. Convective heat transfer loss to the environment from dA_1 is twice as much as the radiation heat transfer loss to the environment. Therefore 66% of the maximum required view factor from the thermal radiation heater is needed to compensate for the convective heat transfer loss from dA_1 to the environment and 34% is needed to compensate for the thermal radiation heat transfer loss from dA_1 to the environment.

Maximum required view factors, $F_{A2\rightarrow dA1}$, from the thermal radiation heater to the elemental surface, dA_1, temperature show almost a linear relationship. However, the sensitivity of $F_{A2\rightarrow dA1}$ to dA_1 surface temperature increases to the order of 10^{-6} with increasing dA_1 surface temperatures as shown in Figure 4-5.

X and Y location limits where dA_1 can be located for a desired surface temperature at the Z=0 surface are shown in Figure 4-6. Y>5.0 meters and any X at Z=0 is a good location for dA_1 when required $dA_1 < 40\ C$. Y>4.0 meters and any X at Z=0 is a good location for dA_1 when required $dA_1 < 50\ C$.

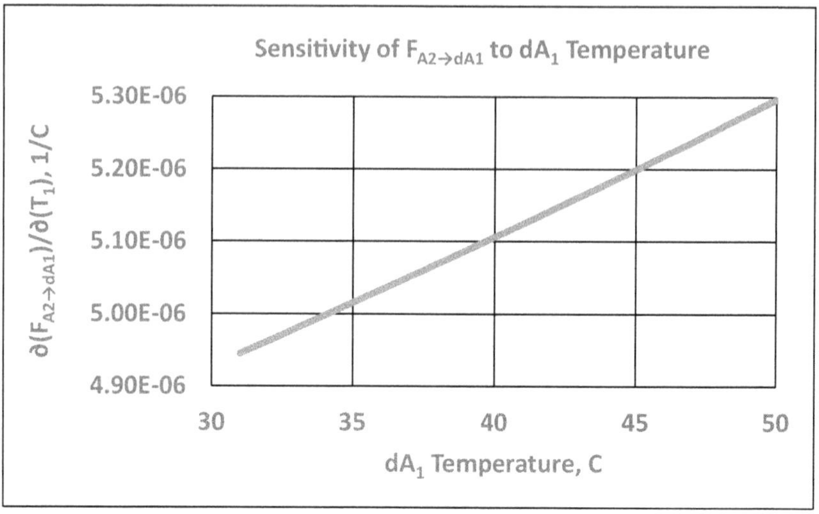

Figure 4-5 Sensitivity of maximum view factors, $F_{A2\to dA1}$, to desired dA_1 surface temperature versus desired dA_1 surface temperature

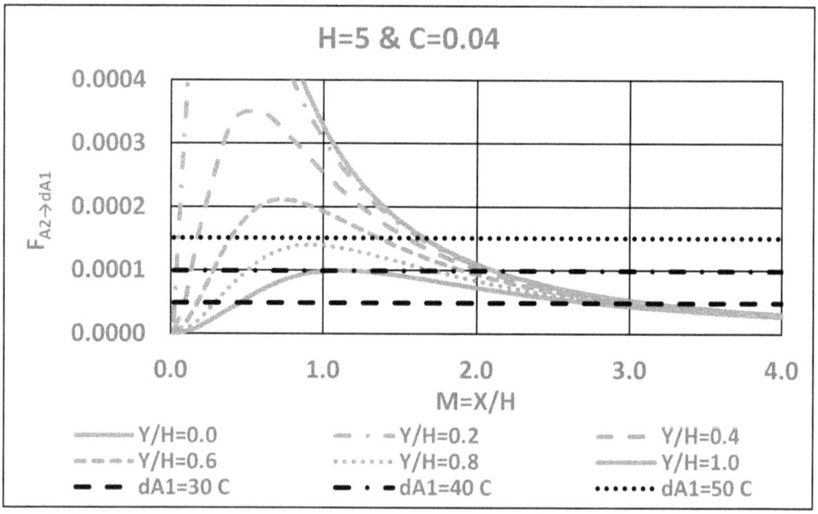

Figure 4-6 View factors $F_{A2\to dA1}$ as a function of $M = \dfrac{X}{H}$ for different values of $N = \dfrac{Y}{H}$ and for elemental surface, dA_1, required temperatures of 30, 40 and 50 C

5

Center of Glass Heat Transfer in Vertical Double-Pane Windows

In this chapter, we will investigate center of glass heat transfer in vertical double pane windows. For contributions to total heat transfer from other parts of a window system such as edges of glass and window frames, see Ref [1]. It is assumed that vertical double panes are parallel, air tight and with no moisture condensation. Also, solar radiation strikes the panes at near normal incidence. In the present analysis, it will be assumed that the outer and the inner surfaces of the glass layer has the same temperature and they are isothermal, since the glass layers are thin, i.e. about 0.003 m thickness, and the thermal conductivity through the glass layer, i.e. about 300 W/(m^2-K), which is two orders of magnitude higher than the overall conductivity through a double pane window, i.e. about 3 W/(m^2-K). Steady-state heat transfer conditions are considered and heat flow is perpendicular to glass pane surfaces and it is one-dimensional. Radiation heat transfer ignores gas absorption between panes. Convective heat transfer assumes high aspect ratio panes. With above assumptions, center of glass heat transfer for a vertical double pane window is illustrated in Figure 5-1.

Using Figure 5-1, heat transfer equations can be written for the unknown outer pane temperature, T_1, and for the unknown inner pane temperature, T_2, as presented in Equations 5-1a and 5-1b, respectively.

$$\left(hc_{1-out} + hr_{1-out}\right)\times\left(T_{out} - T_1\right) + \left(hc_{1-2} + hr_{1-2}\right)\times\left(T_2 - T_1\right) + \alpha_1 \times I_s = 0 \qquad \text{5-1a}$$

$$\left(hc_{1-2} + hr_{1-2}\right)\times\left(T_1 - T_2\right) + \left(hc_{in-2} + hr_{in-2}\right)\times\left(T_{in} - T_2\right) + \alpha_2 \times \tau_1 \times I_s = 0 \qquad \text{5-1b}$$

where $hc_{1\text{-}out}$ in W/(m^2-K), is the forced convection heat transfer coefficient which is determined experimentally, between the outer surface of the outer pane 1 and the windy environmental conditions at the outer surface of the window, $hr_{1\text{-}out}$ in W/(m^2-K), is the radiation heat transfer coefficient between the outside surface of pane 1 and the surroundings and the sky in the infrared spectrum, $hc_{1\text{-}2}$ in W/(m^2-K),

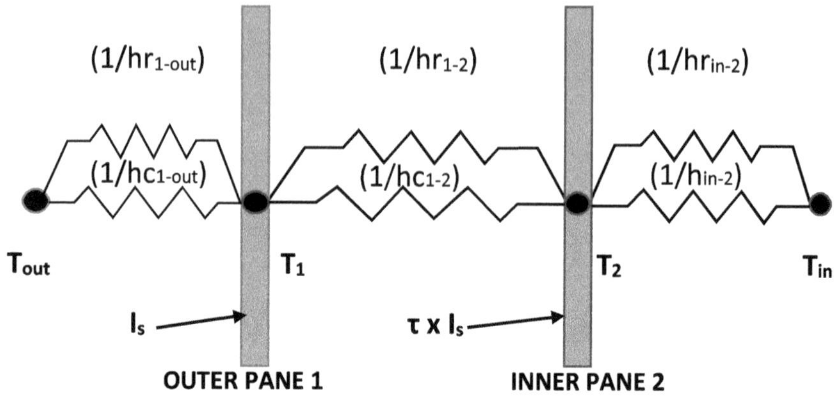

Figure 5-1 Center of glass heat transfer system for a vertical double pane window

is the natural convection heat transfer coefficient which is determined experimentally, between the inner surface of the outer pane 1 and the outer surface of the inner pane 2, $hr_{1\text{-}2}$ in W/(m^2-K), is the radiation heat transfer coefficient between two large parallel planes in the infrared spectrum, namely the inner surface of pane 1 and the outer surface of inner pane 2, $hc_{in\text{-}2}$ in W/(m^2-K), is the natural convection heat transfer coefficient which is determined experimentally, between the inner surface of the inner pane 2 and the room, $hr_{in\text{-}2}$ in W/(m^2-K), is the radiation heat transfer coefficient between the inner surface of pane 2 and the room in the infrared spectrum.

I_s in W/m^2, is the amount of total solar radiation reaching the outer surface of pane 1 both in the visible and in the near infrared wavelength range. I_s includes direct normal irradiation, diffuse irradiation and ground reflected irradiation. In our calculations, I_s has a constant and a typical value since this value depends on several variables such as location, time of day, time of year, cloud coverage, environmental contaminants, and tilt of pane 1 to sun rays. α_1 is the fraction (integrated average over both direction and wavelength) of incident radiation absorbed by pane 1. α_2 is the fraction (integrated average over both direction and wavelength) of incident radiation absorbed by pane 2. τ_1 is the fraction (integrated average over both direction and wavelength) of incident radiation transmitted through pane 1.

The forced convection heat transfer coefficient, $hc_{1\text{-}out}$ for the outer surface of the window and the windy environmental conditions is given as empirical correlations resulting from experiments in Equations 5-2. See Ref [20] for details. On the windward side of the building, the direction of the wind with respect to the azimuth angle of pane 1 has to be less than 90 degrees and on the leeward side of the building it has to be greater than 90 degrees.

$$hc_{1-out} = 8.07 \times V_w^{0.605} \; for \; V_w > 2m/s \; and \; for \; windward \; side \qquad 5\text{-}2a$$

$$hc_{1-out} = 12.27 \; for \; V_w \leq 2m/s \; and \; for \; windward \; side \qquad 5\text{-}2b$$

$$hc_{1-out} = 18.64 \times \left(0.3 + 0.05 \times V_w\right)^{0.605} \; for \; leeward \; side \qquad 5\text{-}2c$$

where V_w is the wind speed in m/s.

The natural convection heat transfer coefficient, $hc_{1\text{-}2}$ through the gap filled with gas such as air, argon or krypton between two vertical panes is given as an empirical correlation resulting from experiments in Equations 5-3. See Ref. [14] for details.

$$Nu = 0.035 \times Ra^{0.38} \; for \; Ra > 6{,}783 \qquad 5\text{-}3a$$

$$Nu = 1 \; for \; Ra \leq 6{,}783 \qquad 5\text{-}3b$$

where *Nu* is the dimensionless Nusselt number which is the ratio of natural convection heat transfer to conduction heat transfer through gas in the glazing gap, namely $Nu = \dfrac{hc_{1-2} \times S}{k}$. *S* is the glazing gap width in meters. *k* is the thermal conductivity of the gas in the glazing gap in W/(m-K). *Ra* is the dimensionless Rayleigh number which correlates experimental results of natural convection heat transfer to Nusselt number. *Ra* is the product of dimensionless Grashof, *Gr*, number which represents the ratio of buoyancy forces acting on a gas to viscous forces and dimensionless Prandtl, *Pr*, number which represents the ratio of kinematic viscosity to thermal diffusivity and they are defined below:

$$Gr = \frac{g\beta\left(T_1 - T_2\right)S^3}{v^2} \qquad Pr = \frac{v}{a} = \frac{(\mu/\rho)}{\left(k/\rho c_p\right)}$$

where $g = 9.81 \; m/s^2$ is the gravitational acceleration, β is the coefficient of thermal expansion of gas in the glazing gap in 1/K, v is the kinematic viscosity of gas in the glazing gap in m^2/s, μ is viscosity of the gas in the glazing gap in $(N\text{-}s)/m^2$, ρ is the density of gas in the glazing gap in kg/m^3, a is the thermal diffusivity of the gas in the glazing gap in m^2/s, and c_p is the specific heat at constant pressure of the gas in the glazing gap in J/(kg-K). During present calculations, above thermo-physical properties of the gas in the glazing gap is taken at average temperature of pane 1 and pane 2, namely at $0.5 \times (T_1 + T_2)$.

The natural convection heat transfer coefficient, $hc_{in\text{-}2}$ between the room air and the inner surface of the vertical pane 2 is given as an empirical correlation resulting from experiments in Equations 5-4. See Ref. [1] for details. This relationship is independent of the height of the vertical pane 2. It is applicable for $-10\;C < T_2 < 55\;C$ and for $3\;C < (T_{in} - T_2) < 30\;C$.

$$hc_{in-2} = 1.77\times\left(T_{in} - T_2\right)^{0.25} \qquad 5\text{-}4$$

Next radiation heat transfer coefficients in Equation-1 are determined. Infrared radiation heat transfer from the outer surface of pane 1 to outside, $QR_{1\text{-}out}$ is the sum of two components as shown in Equation 5-5. One component is the infrared radiation heat transfer to the environment $QR_{1\text{-}env}$ and the other is the infrared radiation heat transfer to the sky $QR_{1\text{-}Sky}$. In the present analysis, emissivity of the front surface of pane 1, $\varepsilon_{1\text{-}f} = 0.85$, and the emissivity of the sky, $\varepsilon_{sky} = 0.75$, are assumed to be constants, namely they are averaged over all directions, wavelengths and applicable temperatures. Also the temperature of the sky is assumed to be the same as the temperature of the outside.

$$QR_{1-out} = QR_{1-env} + QR_{1-sky} = hr_{1-out}\times\left(T_1 - T_{out}\right) \qquad 5\text{-}5$$

$QR_{1\text{-}env}$ is given in Equation 5-6. $F_{1\text{-}Sky}$ is the view factor from the front surface of pane 1 to the sky and it has a value of 0.5 in the present analysis. σ is the Stefan-Boltzmann constant with the value of $5.67\text{x}10^{-8}$ W/(m^2-K^4). A_1 is the outside surface area of pane 1 in m^2.

$$QR_{1-env} = A_1\times\left(1 - F_{1-sky}\right)\times\varepsilon_{1-f}\times\sigma\times\left(T_1^4 - T_{out}^4\right) \qquad 5\text{-}6$$

$QR_{1\text{-}Sky}$ is presented in Equation 5-7. $QR_{1\text{-}Sky}$ is the net infrared radiation heat transfer between the front surface of pane 1 and the sky, namely infrared radiation emission from the front surface of pane 1 minus the infrared irradiation from the sky.

$$QR_{1-sky} = A_1\times F_{1-sky}\times\varepsilon_{1-f}\times\sigma\times\left(T_1^4 - \varepsilon_{sky}\times T_{out}^4\right) \qquad 5\text{-}7$$

Combining Equations 5-5, 5-6 and 5-7, infrared radiation heat transfer coefficient from the outside surface of pane 1 to outside is given in Equation 5-8.

$$hr_{1-out} = \frac{\varepsilon_{1-f}\times\sigma\times\left\{T_1^4 - \left[1 - F_{1-sky}\times\left(1-\varepsilon_{sky}\right)\right]\times T_{out}^4\right\}}{\left(T_1 - T_{out}\right)} \qquad 5\text{-}8$$

In order to evaluate infrared radiation heat transfer between two inner surfaces panes, namely $QR_{1\text{-}2}$, it is assumed that both surfaces are gray and opaque, namely infrared radiation transmittance is zero and sum of emissivity and reflectivity are one for both inner surfaces. Irradiation of gray and opaque two large parallel surfaces are given in Equation 5-9. See Ref. [33] for derivation of Equation 5-9.

$$QR_{1-2} = A_1\times\sigma\times\frac{\left(T_1^4 - T_2^4\right)}{\dfrac{1}{\varepsilon_{1-back}} + \dfrac{1}{\varepsilon_{2-front}} - 1} \qquad 5\text{-}9$$

where $\varepsilon_{1\text{-}back}$ and $\varepsilon_{2\text{-}front}$ are constant emissivities for back surface of pane 1 and front surface of pane 2, respectively. In the present analysis, these emissivities have a value of 0.85 for uncoated surfaces and have a value of 0.1 or 0.2 for thin film coated surfaces. Then the expression for $hr_{1\text{-}2}$ can be obtained as shown in Equation 5-10.

$$hr_{1-2} = \sigma \times \frac{\left(T_1^4 - T_2^4\right)}{\left(\frac{1}{\varepsilon_{1-back}} + \frac{1}{\varepsilon_{2-front}} - 1\right) \times \left(T_1 - T_2\right)} \qquad \text{5-10}$$

$QR_{in\text{-}2}$ can be determined by assuming that the interior walls of the room behaves like a blackbody and $A_2 \ll A_{in}$. Then infrared radiation heat transfer between inner walls of the room and the back surface of pane 2 can be expressed as shown in Equation 5-11. In the present analysis, emissivity of the back surface of pane 2 is a constant, i.e., $\varepsilon_{2\text{-}back} = 0.85$.

$$QR_{in\text{-}2} = A_2 \times \sigma \times \varepsilon_{2-back} \times \left(T_{in}^4 - T_2^4\right) \qquad \text{5-11}$$

Then the expression for $hr_{in\text{-}2}$ can be obtained as shown in Equation 5-12.

$$hr_{in-2} = \frac{\sigma \times \varepsilon_{2-back} \times \left(T_{in}^4 - T_2^4\right)}{\left(T_{in} - T_2\right)} \qquad \text{5-12}$$

Coupled and non-linear Equations 5-1 for center of glass heat transfer in vertical double pane windows can be solved for T_1 and for T_2 iteratively by using Equations 5-2 through 5-12. In present calculations, iteration is stopped when convergence in pane temperatures is less than 0.05 K. All calculations are done for winter conditions, namely $T_{out} = -18\ C$ and $T_{in} = 21\ C$. Also a constant solar radiation reaching the outer surface of pane 1 is assumed, namely $I_s = 500\ W/m^2$. Glass panes had constant solar optical properties, namely $\tau = 0.86$ and $\alpha = 0.05$.

Three different types of fill gases are analyzed between the two panes, i.e. air, argon and krypton. Thermophysical properties of these gases can be found in any heat transfer book, i.e. see Ref. [13].

After T_1 and T_2 are determined, the center of glass overall heat transfer coefficient U in W/(m^2-K) is obtained as a function glazing gap width between the two vertical panes using Equation 5-13.

$$U = \frac{1}{\frac{1}{\left(hc_{1-out} + hr_{1-out}\right)} + \frac{1}{\left(hc_{1-2} + hr_{1-2}\right)} + \frac{1}{\left(hc_{in-2} + hr_{in-2}\right)}} \qquad \text{5-13}$$

Equation 5-13 covers only the heat transfer through the central glazed portion of the vertical double pane window. For total heat transfer coefficient through a vertical double pane window, edge of glass and frame heat transfer should also be considered, i e. see Ref. [1].

U versus glazing gap width is shown in Figure 5-2 for glazing gap filled with air. The upper curve is for a vertical double pane with uncoated inner surfaces. The lower curve is for a vertical double pane whose inner surfaces are coated with low emissivity material such as silicon oxide and silver that are predominantly transparent to visible and to near infrared radiation. In the present case, the emissivity of the coating material is assumed to be 0.2. In the uncoated case, as the glazing gap width increases, the center of glass overall heat transfer coefficient decreases and asymptotes to i.e. 2.71 W/(m^2-K) after a glazing gap width of 0.015 meter. In the coated case, as the glazing gap width increases, the center of glass overall heat transfer coefficient decreases and asymptotes 1.75 W/(m^2-K) after a glazing gap width of 0.013 meter. In the present case, when the inner surfaces of a vertical double pane are coated with low emissivity material, i.e. $\varepsilon_{coating} = 0.2$, there can be a 35.4% reduction in heat transfer through the central glazed portion of a vertical double pane window.

The sensitivities of *U* with respect to glazing gap width are shown in Figure 5-3 for a glazing gap filled with air. The upper curve is for a vertical double pane with uncoated inner surfaces. The lower curve is for a vertical double pane whose inner surfaces are coated with a 0.2 emissivity material.

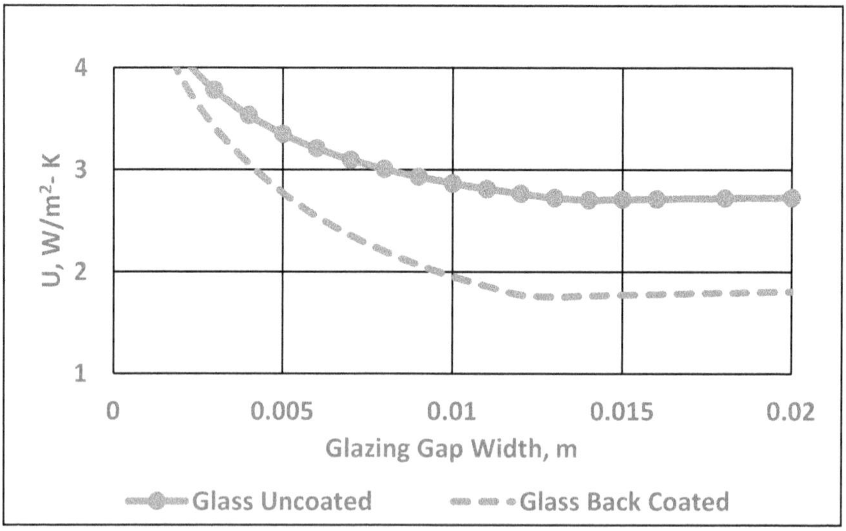

Figure 5-2 U versus glazing gap. Glazing gap filled with air. For coated case, the coating material emissivity is 0.2.

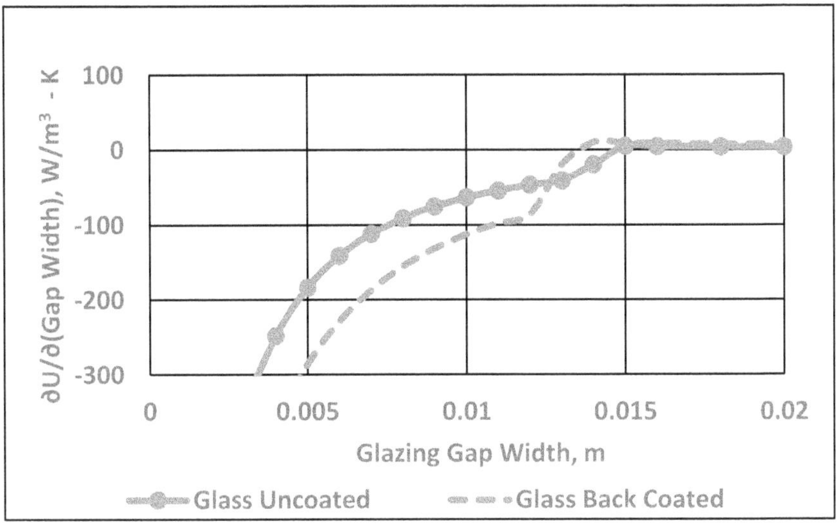

Figure 5-3 Sensitivity of U to glazing gap width. Glazing gap filled with air. For coated case, the coating material emissivity is 0.2.

Absolute value of sensitivity for uncoated inner surfaces starts high at low glazing gap widths and asymptotes to zero at 0.015 meter glazing gap width. Absolute value of sensitivity for coated inner surfaces also starts high at low glazing gap widths and asymptotes to zero at 0.013 meter glazing gap width. During approaching zero sensitivity, the non-smooth behavior of sensitivity curves is due to large glazing gap widths that were taken during calculations, i.e. 0.001 meter, and due to stopping iterations for pane temperatures when convergence is less than 0.05 K.

U versus glazing gap width is shown in Figure 5-4 for glazing gap filled with argon. The upper curve is for a vertical double pane with uncoated inner surfaces. The lower curve is for a vertical double pane whose inner surfaces are coated with low emissivity material, i.e. $\varepsilon_{coating} = 0.2$. In the uncoated case, as the glazing gap width increases, the center of glass overall heat transfer coefficient decreases and asymptotes to 2.57 W/(m^2-K) which is 5.2% lower than the air filled glazing gap case, after a glazing gap width of 0.013 meter. In the coated case, as the glazing gap width increases, the center of glass overall heat transfer coefficient decreases and asymptotes to 1.48 W/(m^2-K) after a glazing gap width of 0.012 meter. In the present case, when the inner surfaces of a vertical double pane are coated with low emissivity material, i.e. $\varepsilon_{coating} = 0.2$, there can be a 42.4% reduction in heat transfer through the central glazed portion of a vertical double pane window.

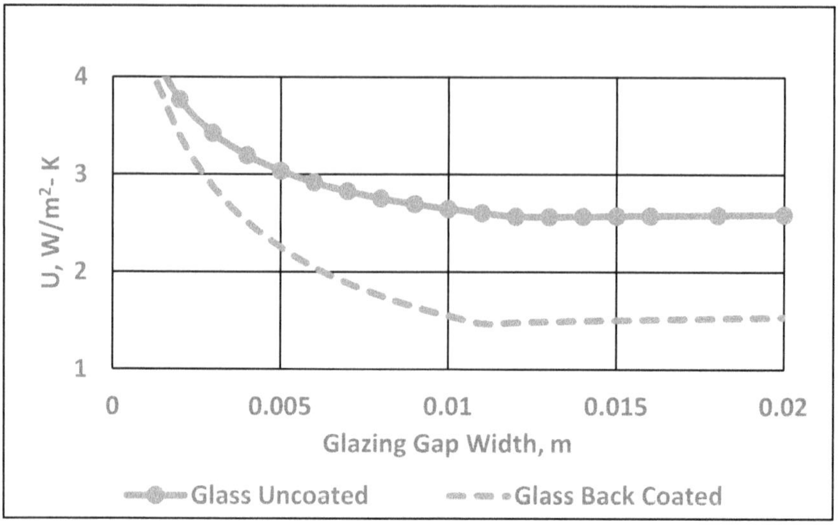

Figure 5-4 U versus glazing gap. Glazing gap filled with argon. For coated case, the coating material emissivity is 0.2.

The sensitivities of U with respect to glazing gap width is shown in Figure 5-5 for glazing gap filled with argon. The upper curve is for a vertical double pane with uncoated inner surfaces. The lower curve is for a vertical double pane whose inner surfaces are coated with a 0.2 emissivity material. Absolute value of sensitivity for uncoated inner surfaces starts high at low glazing gap widths and asymptotes to zero at 0.013 meter glazing gap width. Absolute value of sensitivity for coated inner surfaces also starts high at low glazing gap widths and asymptotes to zero at 0.012 meter glazing gap width. During approaching zero sensitivity, the non-smooth behavior of sensitivity curves is due to large glazing gap widths that were taken during the calculations, i.e. 0.001 meter, and due to stopping iterations for pane temperatures when convergence is less than 0.05 K.

U versus glazing gap width is shown in Figure 5-6 for glazing gap filled with krypton. The upper curve is for a vertical double pane with uncoated inner surfaces. The lower curve is for a vertical double pane whose inner surfaces are coated with low emissivity material, i.e. $\varepsilon_{coating} = 0.2$. In the uncoated case, as the glazing gap width increases, the center of glass overall heat transfer coefficient decreases and asymptotes to 2.56 W/(m^2-K) which is 5.5% lower than the air filled glazing gap case, after a glazing gap width of 0.008 meter. In the coated case, as the glazing gap width increases, the center of glass overall heat transfer coefficient decreases and asymptotes to 1.47 W/(m^2-K) after glazing gap width of 0.008 meter. In the present case, when the inner surfaces of a vertical double pane are coated with low emissivity material, i.e. $\varepsilon_{coating} = 0.2$, there can be a 42.6%

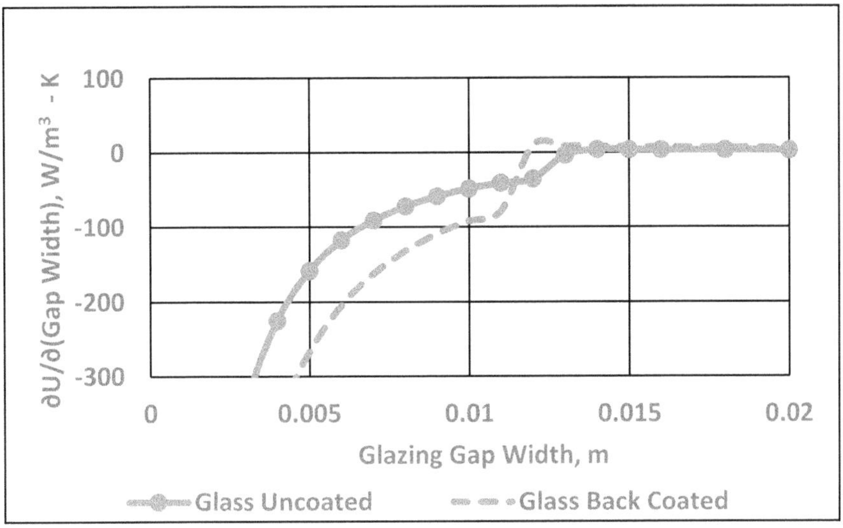

Figure 5-5 Sensitivity of U to glazing gap width. Glazing gap filled with argon. For coated case, the coating material emissivity is 0.2.

reduction in heat transfer through the central glazed portion of a vertical double pane window. Also the glazing gap can be much smaller, i.e. 0.008 meter for krypton filled and coated glazing gap versus 0.013 meter for air filled and coated glazing gap.

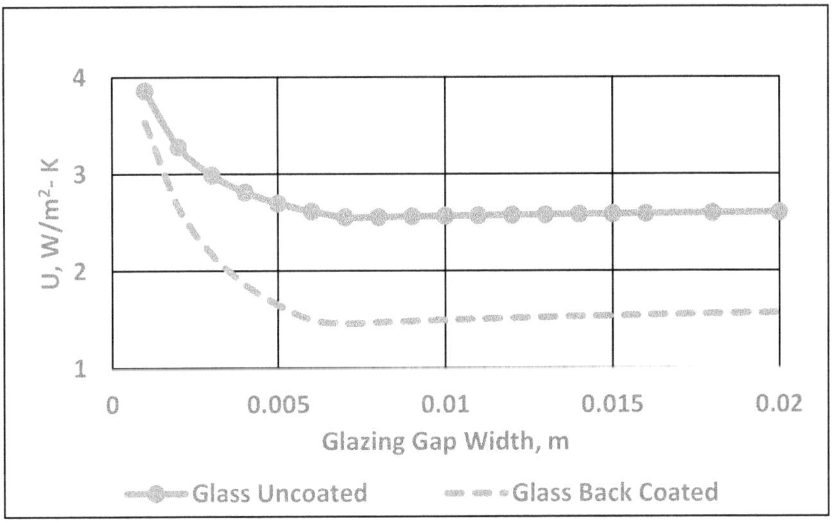

Figure 5-6 U versus glazing gap. Glazing gap filled with krypton. For coated case, the coating material emissivity is 0.2.

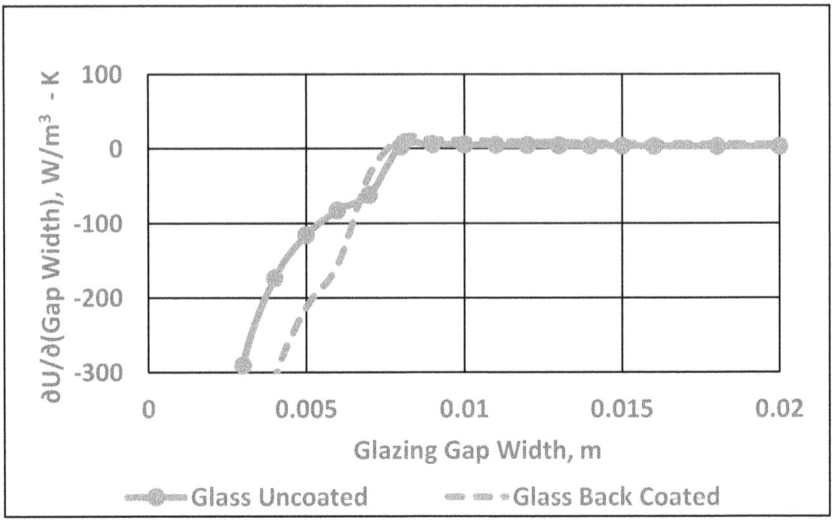

Figure 5-7 Sensitivity of U to glazing gap width. Glazing gap filled with krypton. For coated case, the coating material emissivity is 0.2.

The sensitivity of *U* with respect to glazing gap width is shown in Figure 5-7 for glazing gap filled with krypton. The upper curve is for a vertical double pane window with uncoated inner surfaces. The lower curve is for a vertical double pane window whose inner surfaces are coated with a 0.2 emissivity material. Absolute value of sensitivity for uncoated inner surfaces starts high at low glazing gap widths and asymptotes to zero at 0.008 meter glazing gap width. Absolute value of sensitivity for coated inner surfaces also starts high at low glazing gap widths and asymptotes to zero at 0.008 meter glazing gap width. During approaching zero sensitivity, the non-smooth behavior of sensitivity curves is due to large glazing gap widths that were taken during the calculations, i.e. 0.001 meter, and due to stopping iterations for pane temperatures when convergence is less than 0.05 K.

Figure 5-8 shows the comparison of results from Figures 5-2, 5-4, and 5-6, namely glazing gaps filled with air, argon and krypton. For coated cases, the coating material has an emissivity of 0.2. For vertical double pane windows, the asymptotic value of center of glass overall heat transfer coefficient is lowest for krypton filled gaps when inner pane surfaces coated with low emissivity material.

Filling the glazing gap of a vertical double pane window with krypton which has a low thermal conductivity of 0.091 W/(m-K) at zero temperature and at atmospheric pressure and by coating inner surfaces of panes with low emissivity

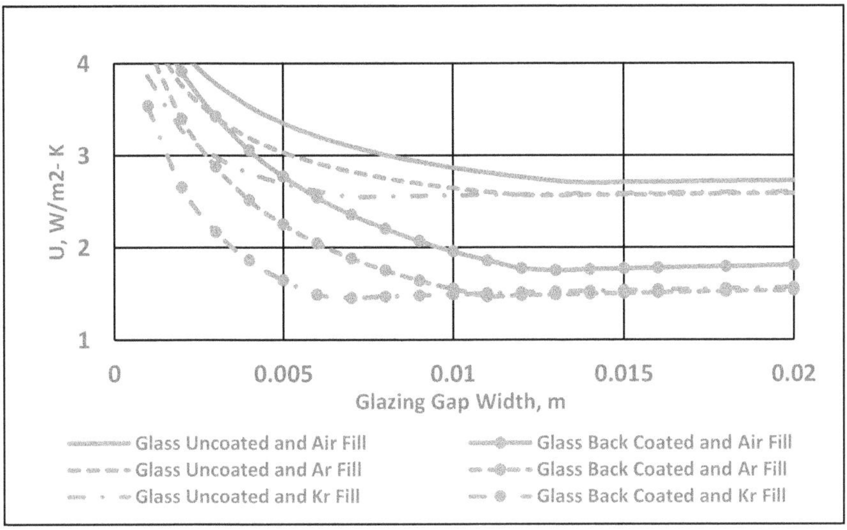

Figure 5-8 U versus glazing gap width. Comparison of gaps filled with air, argon and krypton and comparison of inner pane surfaces uncoated and coated with 0.2 emissivity material

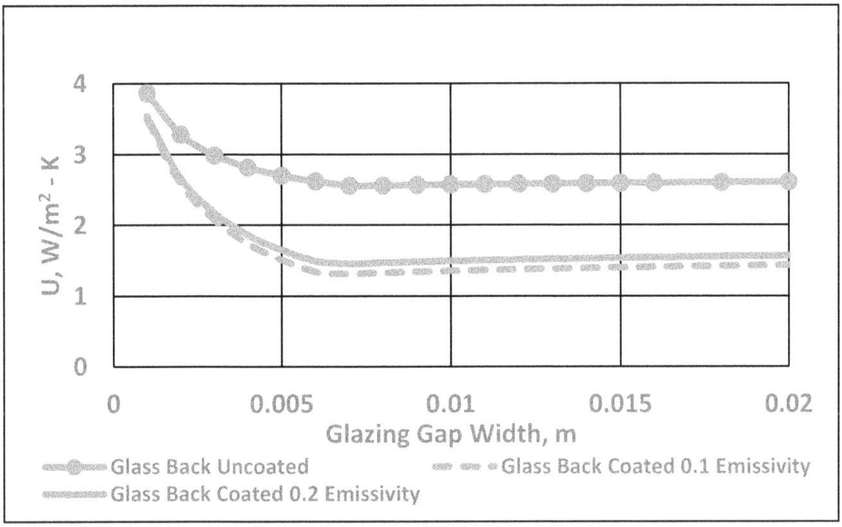

Figure 5-9 U versus glazing gap width. Glazing gap filled with krypton. Comparison of inner pane surfaces uncoated and coated with 0.1 and 0.2 emissivity material

material, i.e. 0.1, provide the lowest center of glass overall heat transfer coefficient as shown in Figure 5-9. If the coating material emissivity goes up to 0.2 from 0.1, the asymptotic value of center of glass overall heat transfer coefficient increases by 9.4%. For uncoated inner surfaces of panes, the asymptotic value of center of glass overall heat transfer coefficient increases by 81.8%. Also, krypton filled gaps can provide the lowest center of glass overall heat transfer coefficient at very low gap glazing widths, i.e. 0.008 meter.

6

Thermal Radiation Effects on Human Skin from a Nuclear Explosion

Approximately 35% of energy released from a nuclear explosion is in the form of thermal radiation. Surface of a nuclear explosion fireball emits large amounts of ultraviolet, visible and infrared rays in the first few seconds after a nuclear explosion. Approximately 50% of energy released from a nuclear explosion is in the form of a pressure wave called a shock wave or it is also called a blast wave for air bursts. This pressure wave develops immediately after the burst and moves outward from the growing fireball. Approximately 5% of energy released from a nuclear explosion is characterized as the initial nuclear radiation that is produced in the first minute of the explosion and which produces high energy neutrons and gamma rays with wavelengths less than 10^{-4} µm. Approximately 10% of energy released from a nuclear explosion is characterized as the residual nuclear radiation that emits gamma rays and beta particles due to delayed radioactivity of explosion residues.

In this chapter, thermal energy that radiates from the surface of a fireball resulting in a nuclear explosion will be investigated. Only nuclear explosions (bursts) occurring in air above the surface of the Earth and below five kilometers altitude will be considered. See Ref. [9] for details. In literature, experimental results of thermal radiation power of a nuclear fireball is expressed by normalizing it to a maximum power, P_{max}, versus time after explosion normalized to time at maximum power, t_{max}. See Ref. [9] for details. The standard measurement for power yielding from a nuclear explosion, Y, is expressed in kilotons (kT), namely one kiloton of power from a nuclear explosion yields same amount of energy as in explosion of 1,000 tons of TNT (trinitrotoluene).

In a nuclear explosion about 80% of thermal energy is released from the surface of the fireball within about first ten seconds after the burst. At the maximum thermal power point after the burst, namely at the second radiation pulse, the maximum temperature of the fireball surface is about 7,400 K as compared to the average surface temperature of our Sun which is about 5,800 K. Empirical results for maximum thermal radiation power, P_{max}, and time at maximum thermal radiation power, t_{max}, for an air burst below five kilometers altitude are given in Equations 6-1 and 6-2, respectively. See Ref. [9] for details.

$$P_{max} = 3.18\times Y^{0.56} kT/\sec \quad 6\text{-}1$$

$$t_{max} = 0.0417\times Y^{0.44}\sec \quad 6\text{-}2$$

However, since 1 $kT = 4.18 \times 10^{12}$ *Joules*, Equation 6-1 can be rewritten as shown in Equation 6-3.

$$P_{max} = 1.33 \times 10^{13} \times Y^{0.56}\text{ Watts} \quad 6\text{-}3$$

Maximum value of thermal radiation power, P_{max}, versus nuclear explosion yield, Y, is shown in Figure 6-1. When nuclear explosion yield increases 100 times, the maximum value of thermal radiation power only increases 13.2 times. This shows that most of the burst energy is consumed inside the fireball and does not reach its surface as the fireball diameter gets large with increasing nuclear explosion yield. Change in maximum value of thermal radiation power, dP_{max}, to change in nuclear explosion yield, dY, versus nuclear explosion yield, Y, is shown in Figure 6-2. Change in maximum value of thermal radiation power, dP_{max}, to change in nuclear explosion yield, dY, is proportional to $1/t_{max}$. This change decreases as the nuclear explosion yield increases and as the fireball diameter increases.

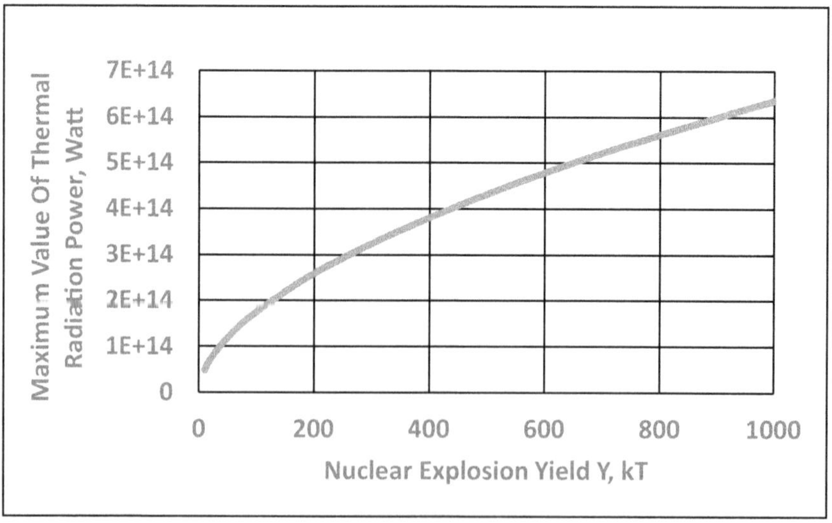

Figure 6-1 Maximum value of thermal radiation power, P_{max}, versus nuclear explosion yield, Y

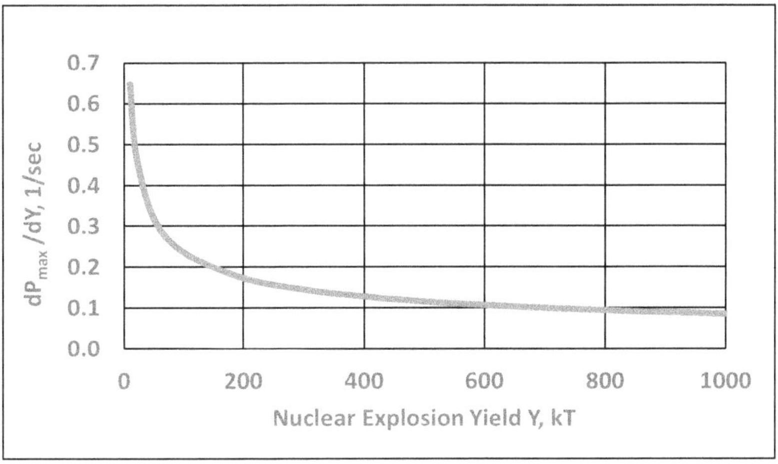

Figure 6-2 Change in maximum value of thermal radiation power, dP_{max}, to change in nuclear explosion yield, dY, versus nuclear explosion yield, Y

Time, t_{max}, when the maximum thermal radiation power occurs after a blast versus nuclear explosion yield is shown in Figure 6-3. t_{max} increases to 0.44 power of nuclear explosion yield for an air burst as given in Equation 6-2. Change in time when the maximum thermal radiation power occurs, dt_{max}, to change in nuclear explosion yield, dY, versus nuclear explosion yield, Y, is presented in Figure 6-4. $\frac{dt_{max}}{dy}$ decreases to $Y^{-0.56}$ as the nuclear explosion yield increases.

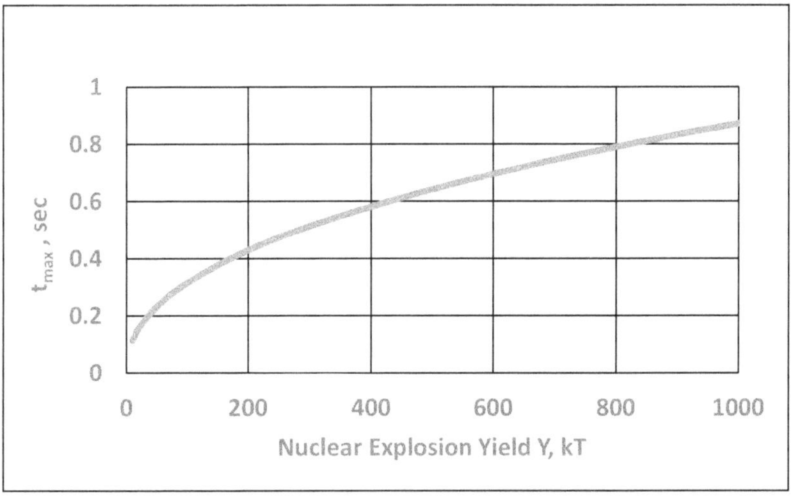

Figure 6-3 Time, t_{max}, when the maximum thermal radiation power occurs after a blast versus nuclear explosion yield, Y

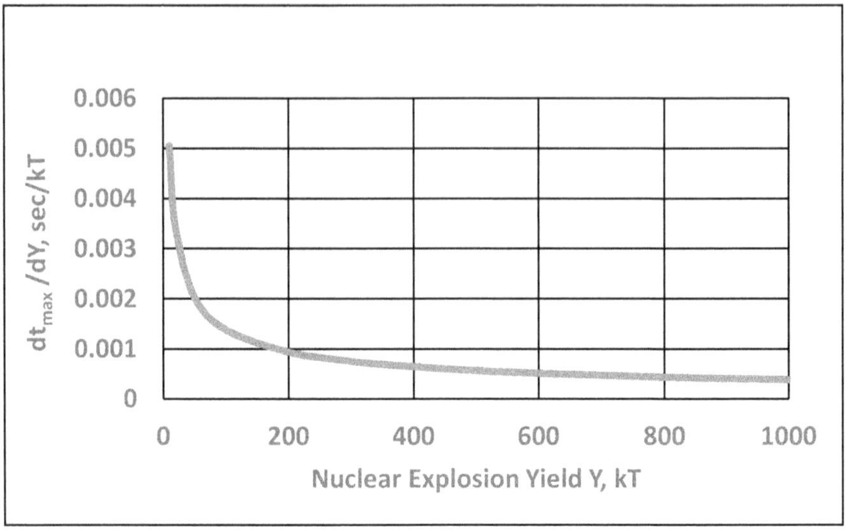

Figure 6-4 Change in time when the maximum thermal radiation power occurs, dt_{max}, to change in nuclear explosion yield, dY, versus nuclear explosion yield, Y

Emission of thermal radiation energy from the surface of a fireball can be estimated by assuming the fireball's surface acts as a blackbody. Total emissive thermal radiation power of a blackbody can be obtained by performing the integration of Planck's spectral distribution function for a blackbody radiation from zero to infinity for all wavelengths. See Equations 1-1 and 1-2 in Chapter 1. Thermal radiation power generated at the blackbody surface of a fireball can be written as shown in Equation 6-4.

$$P = 7.125 \times 10^{-7} \times T^4 \times R^2 \text{ Watts} \quad 6\text{-}4$$

where T is the surface temperature of fireball surface in Kelvin and R is the radius of fireball in meters. It is very complicated experimental and theoretical process to determine the temperature and the radius of a fireball as a function of time after a nuclear explosion due to interacting radiation and hydrodynamic variables. Therefore, thermal radiation power generated at the surface of a fireball versus time after the burst has been obtained by a combination of theory and experiments. Normalized thermal radiation power and percent emitted thermal radiation energy from the surface of a fireball versus normalized time for an air burst is presented in Figure 6-5. See Ref. [9], Chapter 7, for more details. Using this normalized Figure 6-5, thermal radiation power generated by a nuclear explosion's fireball can be calculated at any time after the burst. Again, using this normalized Figure 6-5, percent thermal radiation energy released by a nuclear explosion's fireball can be determined at any time after the burst.

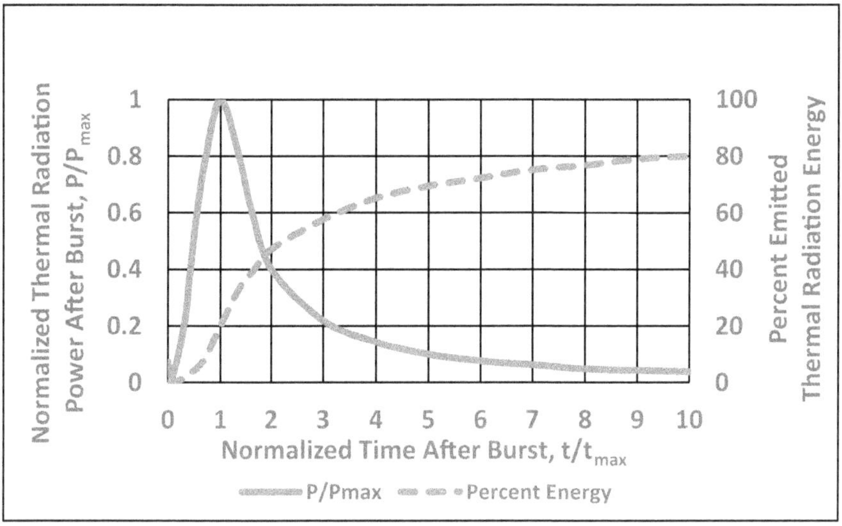

Figure 6-5 Normalized thermal radiation power and percent emitted thermal radiation energy versus normalized time for an air burst

Let us now investigate the thermal radiation exposure of a person standing outdoors at sea level in open surroundings and on a flat terrain at a certain distance X from a nuclear explosion's (air burst's) ground zero location. The intensity of thermal radiation decreases as the inverse square of the distance L from the air burst. L is defined as $\sqrt{X^2 + Z^2}$ where Z is the vertical distance from ground zero point to the air burst point in the sky. It is assumed that thermal radiation from the fireball's surface expands uniformly over a spherical surface with an area of $4 \times \pi \times L^2$.

Also, the intensity of thermal radiation is reduced by absorption and scattering due to air molecules and particles in the atmosphere. Thermal radiation exposure from an air burst is formulated empirically and it is given in Equation 6-5. See Ref. [3] for details.

$$Q = \frac{10^{12} \times 0.35 \times Y \times \tau}{4 \times \pi \times L^2} \qquad \text{6-5}$$

where $0.35 \times Y$ is the thermal radiation fraction of the nuclear explosion yield in kilotons. 10^{12} is the conversion factor from kilotons to calories. τ is an empirical and a non-dimensional factor for visibility which is also called transmittance and it takes into account scattering, absorption and distance covered by thermal radiation from the surface of a fireball. $4 \times n \times L^2$ is the surface area of the uniformly expanding thermal radiation sphere where L is in centimeters.

The transmittance τ is also formulated empirically and it is shown as an exponential decaying function of the distance, L, to the point of burst divided by the decaying constant of visibility, V, as shown in Equation 6-6. See Ref. [3] for details.

$$\tau = \left[1 + 1.4 \times \left(\frac{L}{V} \right) \right] \times exp \left[-2 \times \left(\frac{L}{V} \right) \right] \qquad 6\text{-}6$$

Transmittance is plotted as a function of distance from the point of burst for different visibilities in Figure 6-6. For low visibility, i.e. $V = 10\ km$, 40% of thermal radiation is attenuated in the atmosphere in about 5 km from the point of burst. For high visibility, i.e. $V = 50\ km$, 40% of thermal radiation is attenuated in the atmosphere in about 25 km.

Using Equations 6-5 and 6-6, thermal radiation exposure in calories/cm^2 can be calculated for different distances from the point of air burst and for different visibilities. Figure 6-7 shows thermal radiation exposure in calories/cm^2 for different distances from the point of air burst for 50 km visibility and for different nuclear explosion yields. Thermal radiation exposure attenuates fast for low nuclear explosion yields. Even the 1,000 kT nuclear explosion yield exposure attenuates to 100 calories/cm^2 after 5 kilometers from the point of air burst.

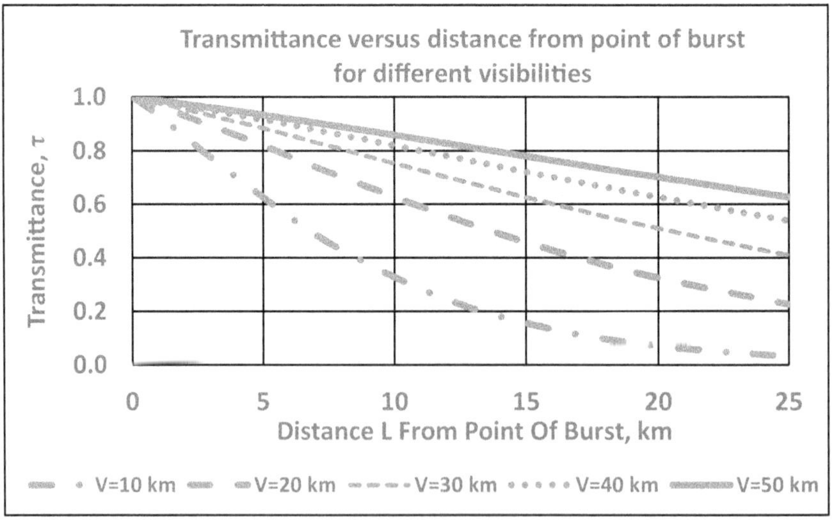

Figure 6-6 Transmittance versus distance from the point of burst for different visibilities

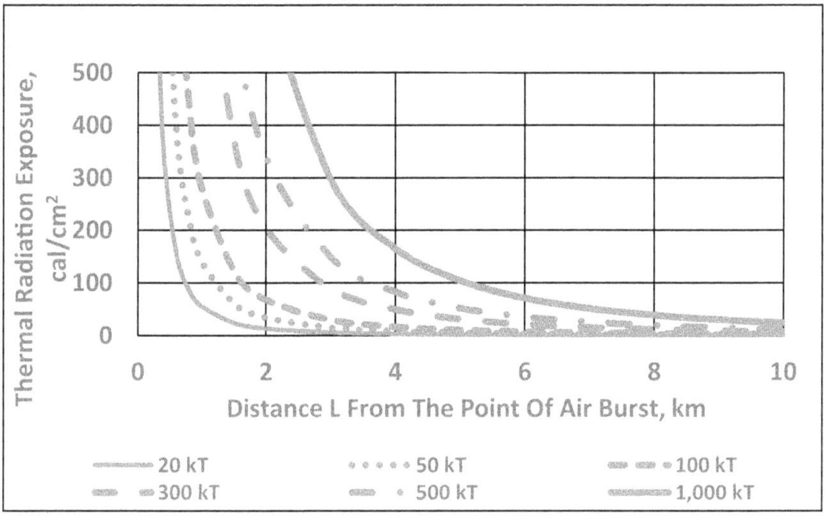

Figure 6-7 Thermal radiation exposure versus distance from the point of burst with 50 km visibility

A person can have first degree burns on his or her skin and tissues below it when thermal radiation exposure is about 3 calories/cm^2, i.e similar to a sunburn. If the thermal radiation exposure from a nuclear explosion increases to about 5 calories/cm^2, second degree burn stage is reached and 25% of body skin and tissues below it can be damaged irreversibly. If the thermal radiation exposure increases to about 8 calories/cm^2, third degree burn stage is reached and 50% of body skin and tissues below it can be damaged and the result can be fatal. Low thermal radiation exposure section, i.e. below 10 calories/cm^2, from Figure 6-7 is shown in Figure 6-8. It can be seen from Figure 6.8 that a third degree burn can be avoided, if a person is away greater than 2.6 km from the point of an air burst of 20 kT nuclear explosion when he or she is in a 50 km visibility environment. Similarly, a third degree burn can be avoided, if a person is away greater than 16.2 km from the point of an air burst of a 1,000 kT nuclear explosion when he or she is in a 50 km visibility environment.

Sensitivities of thermal radiation exposure curves to distance from the point of burst for 50 km visibility and for different nuclear explosion yields are presented in Figure 6-9. These sensitivities are substantial and their absolute values are decreasing toward zero fast. Sensitivities of thermal radiation exposure curves to distance flatten out after a distance of about 10 km and approach zero fast.

Thermal radiation low level exposures versus distance from the point of burst with different visibilities for a 1,000 kT nuclear explosion yield are shown in Figure 6-10. For a 10 km visibility, a third degree burn thermal radiation exposure, namely 8 calories/cm^2 can occur up to 10.4 km distances from the point of burst.

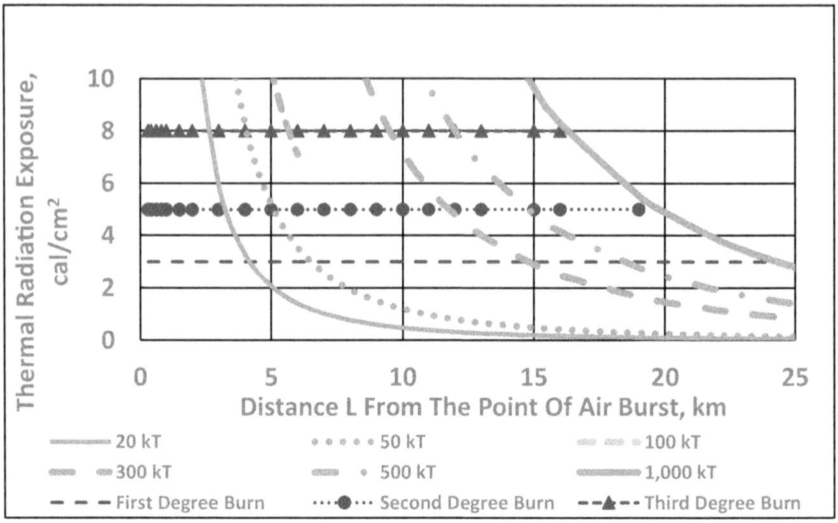

Figure 6-8 Thermal radiation exposure versus distance from the point of burst with 50 km visibility and for different nuclear explosion yields. First degree, second degree and third degree burns lines for human skin and tissues below it are also shown

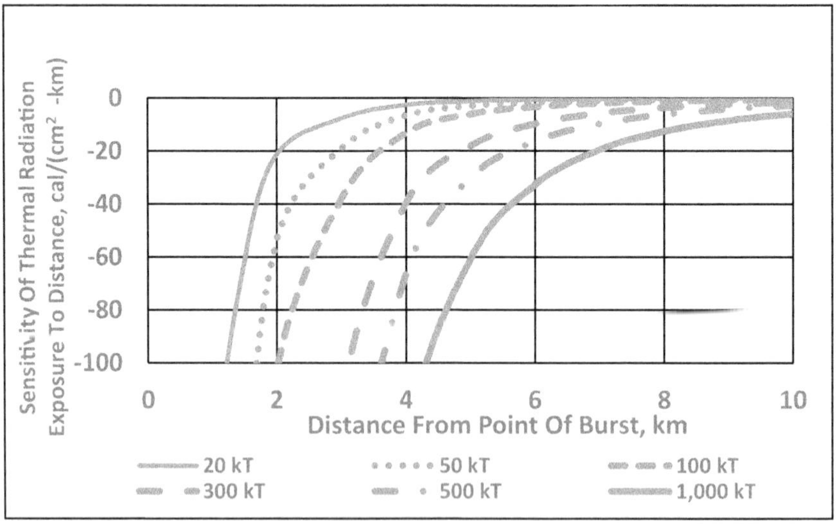

Figure 6-9 Sensitivity of thermal radiation exposure to distance from the point of burst for 50 km visibility and for different nuclear explosion yields

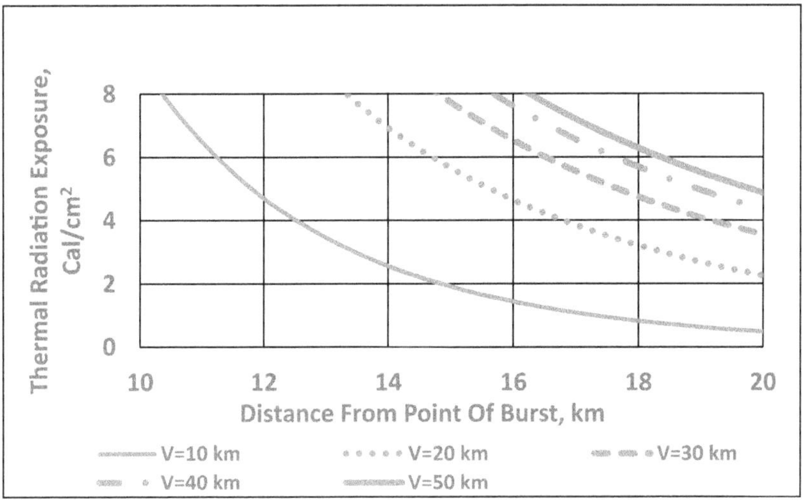

Figure 6-10 Thermal radiation exposures versus distance from the point of burst with different visibilities for a 1,000 kT nuclear explosion yield

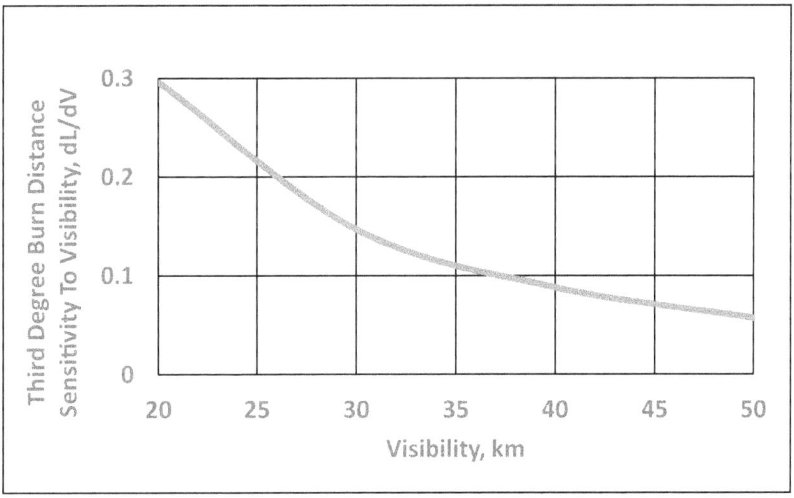

Figure 6-11 Third degree burn distance sensitivity to visibility versus visibility for a 1,000 kT nuclear explosion yield

However, for a 50 km visibility, a third degree burn thermal radiation exposure can occur up to 16.3 km distances from the point of burst.

Third degree burn distance sensitivity to visibility versus visibility for a 1,000 kT nuclear explosion yield is depicted in Figure 6-11. This sensitivity decreases as the visibility increases. The sensitivity decrease is about six-fold from 20 km visibility to 50 km visibility.

7

Exposure to Solar Ultraviolet Radiation on Earth

About 95 % of total solar radiation reaching Earth's surface comprises of visible (380 to 750 nanometers, nm wavelengths) and infrared (700 to $1x10^6$ nm wavelengths) regions of the electromagnetic radiation spectrum. Remaining 5 % of total solar radiation reaching Earth's surface comprises of ultraviolet (UV) (100 to 400 nm wavelengths) region of the electromagnetic radiation spectrum. The UV region of the electromagnetic radiation spectrum is divided into three bands, namely UVA band between 315 and 400 nm wavelengths, UVB band between 280 and 315 nm wavelengths and UVC band between 100 and 280 nm wavelengths. Excessive exposure to solar UV radiation can cause very serious health effects on our skin, i.e. skin cancer, to our eyes, i.e. cataracts, and to our immune system. Also, exposure to solar UV radiation has cumulative adverse effect on human health.

In this chapter, UV radiation on Earth's surface that effects the human skin, namely UV induced erythema on the human skin, will be analyzed using spectrally resolved spectroradiometer data taken at a tropical location, i.e. 19.33 degrees South, and at a location close to the polar circle, i.e. 64.817 degrees North. UV radiation data will be reduced to standard UV index (UVI) formulation as defined by World Health Organization and by ISO Standard 17166. See Ref. [15]. UVI definition is provided in Equation 7-1.

$$UVI = k_{er} \times \int_{290}^{400} E_{\lambda} \times S_{er}(\lambda) \times d\lambda \qquad \text{7-1}$$

Data below 290 nm wavelength is not considered in the present analysis, since UV radiation below 290 nm does not penetrate our Earth's atmosphere due to the ozone layer in the stratosphere and due to pollutants in the troposphere. In Equation 7-1, UVI is a dimensionless number. However, the broadcasted and reported UVI is rounded to the nearest whole number. k_{er} is a constant with a value of 40 m^2/W. E_{λ} is solar spectral irradiance data in $W/(m^2\text{-}nm)$. $d\lambda$ is the wavelength interval in nm that is used during the spectral irradiation data's integration process. $S_{er}(\lambda)$ is a weighting function or an action spectrum for UV induced erythema on human skin. See Ref. [21]. $S_{er}(\lambda)$ is given in Equations 7-2a to 7-2d and plotted in Figure 7-1. Different weighing functions are used in literature for different UV radiation induced damage on humans.

$$S_{er}(\lambda) = 1 \, for \; 250 < \lambda \leq 298 \tag{7-2a}$$

$$S_{er}(\lambda) = 10^{0.094\times(298-\lambda)} \; for \; 298 < \lambda \leq 328 \tag{7-2b}$$

$$S_{er}(\lambda) = 10^{0.015\times(139-\lambda)} \; for \; 328 < \lambda \leq 400 \tag{7-2c}$$

$$S_{er}(\lambda) = 0 \, for \; 400 < \lambda \tag{7-2d}$$

Solar UV irradiance spectral measurements on Earth's surface are affected by solar zenith angle, latitude, cloud cover, altitude, ozone layer thickness and ground reflections. In the present analysis, solar UV irradiance measurement data under clear sky conditions and for different solar zenith angles (time of day) are utilized.

One set of data is from Townsville, Australia (19.33 degrees South, 146.76 degrees East and 30 m. above sea level). See Ref. [2]. Townsville data was recorded middle of summer on 11 January 1996. Townsville data is presented in Figure 7-2. Solar spectral irradiance versus wavelength (290 nm to 400 nm) are shown for different solar zenith angles or times of day. In the present analysis, data is assumed to be symmetric about 12:00 noon.

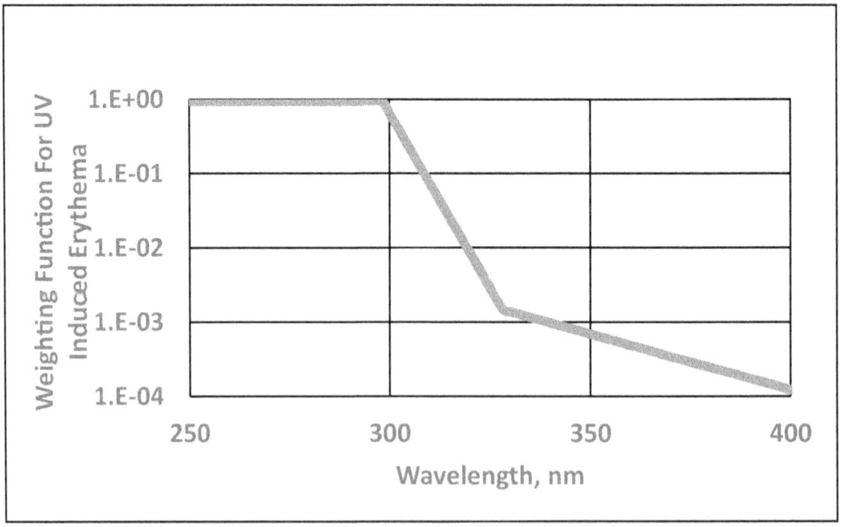

Figure 7-1 Weighting function for UV induced erythema on human skin

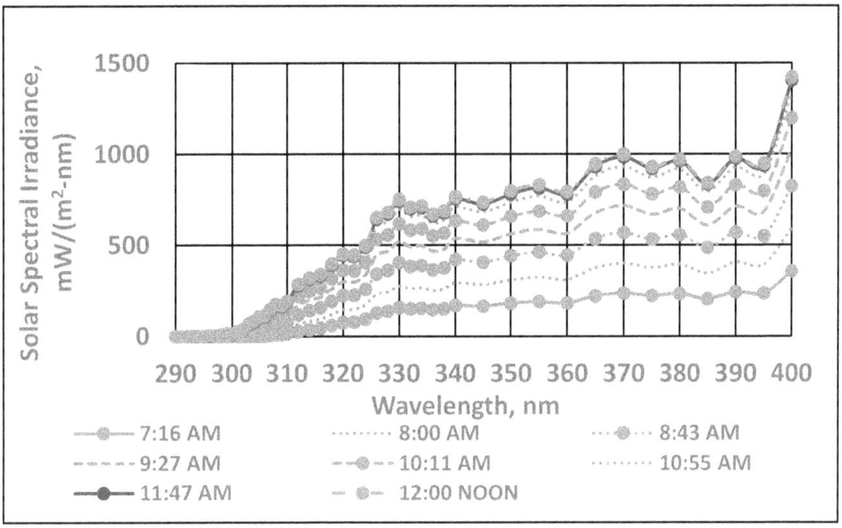

Figure 7-2 Townsville's solar spectral irradiance data on 11 January 1996 under clear skies

Equation 7-1 is integrated for wavelengths from 290 nm to 400 nm with $d\lambda$ varying between 1 to 5 nm in order to obtain UVI for Townsville, Australia on 11 January 1996. Solar spectral irradiance data is assumed to be symmetrical about 12:00 noon. Then the UVI data is fitted with the normal distribution function which is bell shaped and it is perfectly symmetrical about 12:00 noon. The coefficient of determination between data and the normal distribution function is 0.9994. UVI data and the normal distribution function are presented in Figure 7-3.

Table 7-1 Grouping of UVI values and UVI protection requirements

Exposure Category	UVI Value	UVI Protection Requirements
Low	0, 1, 2	No Protection Required
Moderate	3, 4, 5	Protection Required Seek Shade, Wear Shirt And Hat, Put On Sunscreen And Sunglasses
High	6, 7	
Very High	8, 9, 10	Extreme Protection Avoid Being Outside, Seek Shade, Shirt, Hat, Sunscreen And Sunglasses Are A Must
Extreme	11 And Above	

UVI values are evaluated according to the exposure categories specified in Ref. [31] as shown in Table 7-1. Table 7-1 also specifies UVI protection requirements.

UVI versus time of day for Townsville, Australia on 11 January 1996 can be broken down as follows:

Low UVI exposure, i.e. $UVI < 2.5$, occurred in the morning up to 7:39 AM and in the afternoon after 16:21 PM.

Moderate UVI exposure, i.e $2.5 < UVI < 5.5$, occurred in the morning between 7:39 AM and 8:42 AM and in the afternoon between 15:18 PM and 16:21 PM.

High UVI exposure, i.e $5.5 < UVI < 7.5$, occurred in the morning between 8:42 AM and 9:11 AM and in the afternoon between 14:49 PM and 15:18 PM.

Very high UVI exposure, i.e $7.5 < UVI < 10.5$, occurred in the morning between 9:11 AM and 9:51 AM and in the afternoon between 14:09 PM and 14:49 PM.

Extreme UVI exposure, i.e $10.5 < UVI$, started in the morning at 9:51 AM and ended in the afternoon at 14:09 PM. The maximum UVI is 17 at noon. So, we have to avoid being outside for 4 hours during midday, on a clear day during the summer, in Townsville, Australia as shown in Figure 7-3.

The sensitivity of UVI to time of day versus time of day for Townsville, Australia is shown in Figure 7-4. Sensitivity starts at zero and increases to a maximum of 4.62 (1/hr) around 10:19 AM. Then sensitivity starts to decrease and goes through zero at 12:00 noon. The sensitivity is negative after 12:00 noon and goes through a minimum at 14:54 PM. The absolute value of the minimum sensitivity value is again 4.62 (1/hr). Then the sensitivity approaches zero as evening comes.

Let us now compare UVI results obtained in a tropical location above under clear skies to UVI results obtained at a location way up north again under clear skies. The solar spectral irradiation data is taken in Fairbanks, Alaska on 1 June 2019. Fairbanks, Alaska is located at 64.817 North latitude, -147.8792 West longitude. Her elevation is 146 meters. See Ref. [5]. Fairbanks solar irradiation data is provided from 300 nm to 368 nm. UVI is determined using Equations 7-1 and 7-2, namely the weighting function is again for UV induced erythema on human skin.

Then the UVI data is fitted with the normal distribution function which is bell shaped and it is perfectly symmetrical about 12:30 PM. Fairbank's maximum UVI value reaches 5 at 12:30 PM, namely a moderate UVI exposure value, as shown in Figure 7-5. Townsville and Fairbanks UVI distributions during the day are also compared in Figure 7-5.

Low UVI exposure in Fairbanks, Alaska, i.e. $UVI < 2.5$, occurred in the morning up to 8:20 AM and in the afternoon after 16:40 PM.

Moderate UVI exposure in Fairbanks, Alaska, i.e. $2.5 \leq UVI < 5.5$, occurred between 8:20 AM and 16:40 PM.

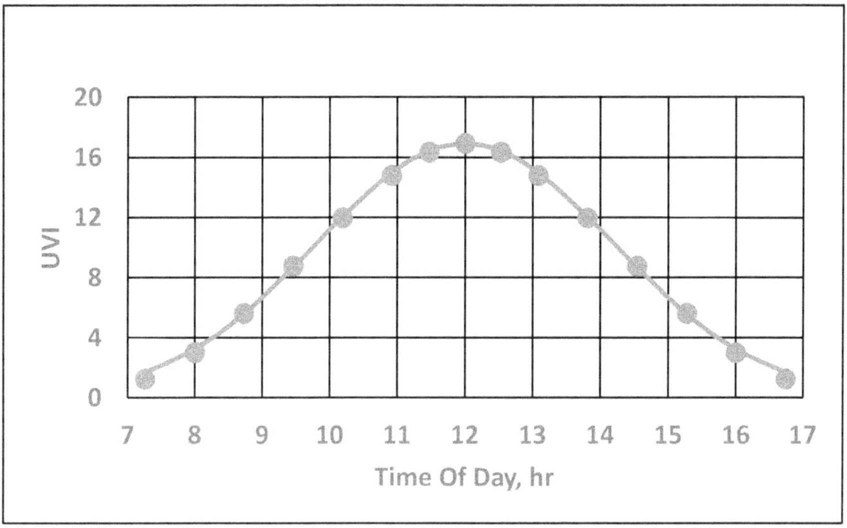

Figure 7-3 UVI versus time of day for Townsville, Australia on 11 January 1996

Comparison of sensitivities of UVI to time of day versus time of day for Townsville, Australia and for Fairbanks, Alaska during a summer day under clear skies is presented in Figure 7-6.

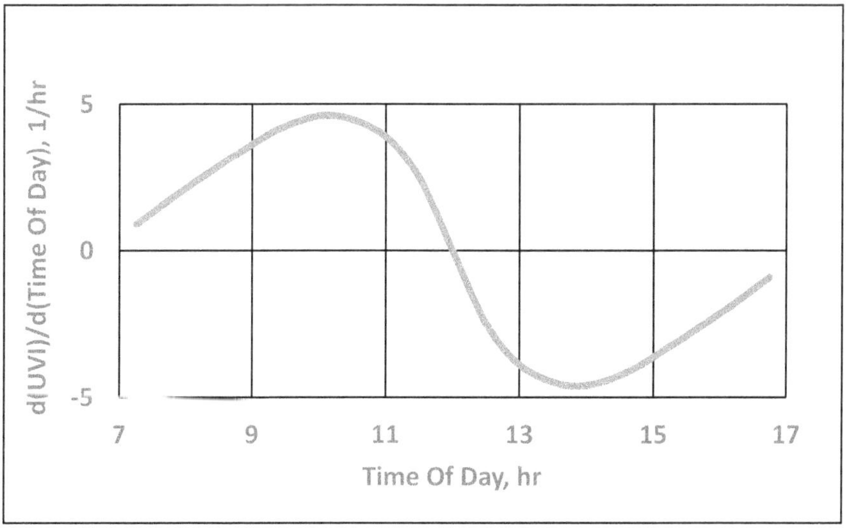

Figure 7-4 Sensitivity of UVI to time of day versus time of day for Townsville, Australia on 11 January 1996

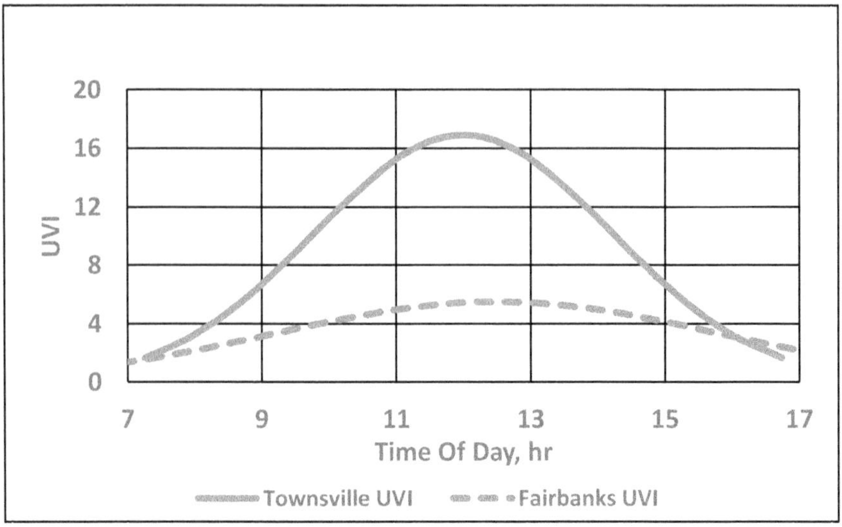

Figure 7-5 Comparison of UVI data for Townsville, Australia and Fairbanks, Alaska during a summer day under clear skies

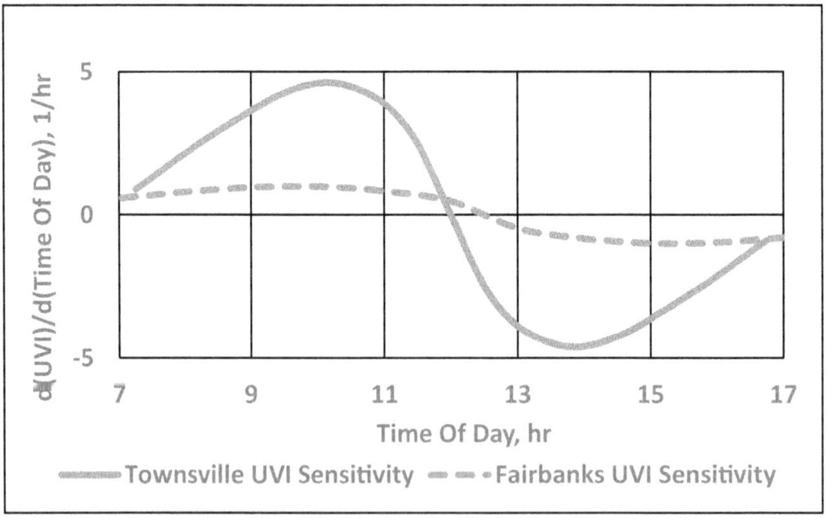

Figure 7-6 Comparison of sensitivities of UVI to time of day versus time of day for Townsville, Australia and for Fairbanks, Alaska during a summer day under clear skies

Sensitivity for Fairbanks, Alaska starts at zero and increases to a maximum of 0.99 (1/hr) around 10:00 AM. Then sensitivity starts to decrease and goes through zero at 12:30 PM. The sensitivity is negative after 12:30 PM and goes through a minimum at 15:00 PM. The absolute value of the minimum sensitivity value is again 0.99 (1/hr). Then the sensitivity approaches zero as evening comes. The ratio of Townsville and Fairbanks maximum sensitivities is 4.67 to 1 which is the same as ratio of minimum sensitivities.

Exposure to solar UV radiation on Earth can vary considerable from place to place as we have seen in the present example. Exposure to solar UV radiation is also affected in addition to latitude by cloud cover, altitude, ozone layer thickness and ground reflections. We should listen carefully to the reported UVI values for that day and take our solar UV radiation exposure precautions accordingly in order to avoid skin cancer and cataracts in our eyes. We have to always remember that solar UV radiation exposure damage on our body is cumulative.

8

Exposure to Ultraviolet Radiation from Fluorescent Lamps

Fluorescent lamps for general lighting come in different powers, sizes and phosphors to emit visible radiant energy. A plastic cover called diffuser that is placed over the lamp effectively eliminates all ultraviolet radiation from the lamp. However, if there is no cover over the lamp and the glass envelope of the lamp is defective and has metal impurities in it, the ultraviolet radiation (UVR) emission from the lamp can cause significant cumulative damage to human eyes and skin. Safety standards for lamps and lighting equipment specify that the effective UVR exposure, i.e. spectrally weighted, has to be below 0.001 W/m^2. This level of human exposure to UVR has no photobiological hazard to skin and eyes. ICNIRP, see Ref. [27], standard limits the UVR exposure energy to human skin and eyes at 30 $Joules/m^2$. With this standard UVR exposure energy limit, the maximum effective UVR exposure duration can be calculated using Equation 8-1.

$$t_{maximum\ exposure\ duration} = \frac{30}{E_{effective}},\ seconds \qquad 8\text{-}1$$

where $E_{effective}$ in W/m^2 is the effective UVR power on the exposed human surface and it is defined by the integral Equation 8-2 in the ultraviolet radiation region between 100 nm and 400 nm wavelengths. Mostly wavelengths between 100 to 180 nm are ignored since these electromagnetic waves are absorbed in air.

$$E_{effective} = \int_{100}^{400} E_{\lambda} \times S_{er}(\lambda) \times d\lambda \qquad 8\text{-}2$$

Since the UVR exposure energy should not exceed 30 J/m^2 and also per the photobiological safety standards for lamps and lighting equipment, effective UVR exposure power should be below 0.001 W/m^2, maximum exposure time to this level of lighting can be 8.3 hours per day. In Equation 8-2, E_{λ} is the spectral irradiance at the surface of the exposed human tissue in $W/(m^2\text{-}nm)$. $d\lambda$ is

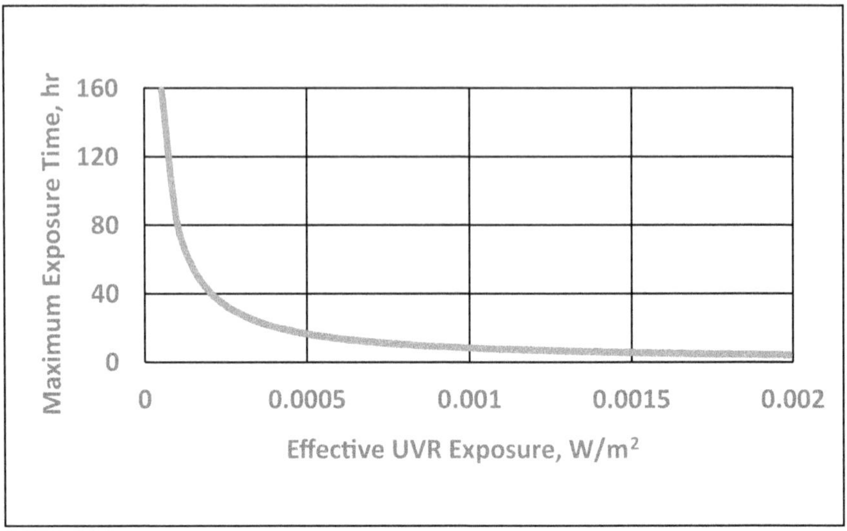

Figure 8-1 Maximum exposure time to reach the standard limit of 30 Joules/m^2 versus effective UVR exposure for lamps and lighting equipment

the wavelength interval in nm that is used during the spectral irradiation data's integration process. $S_{er}(\lambda)$ is a weighting function or an action spectrum for UV induced erythema on human skin. See Ref. [21]. $S_{er}(\lambda)$ is given in Chapter 7 in Equations 7-2a to 7-2d and plotted in Figure 7-1. Different weighing functions are used in literature for different UVR induced damage on humans. Maximum exposure time in hours versus effective UVR exposure in W/m^2 is shown in Figure 8-1 by using the standard UVR energy exposure limit of 30 Joules/m^2 to human skin and eyes.

The solar UV index exposure levels do not even fit into the range of the effective UVR exposure for lamps and lighting equipment in Figure 8-1. For example, a UV index of 10, see Chapter 7 for UV index definition, will have an effective solar UVR exposure of 0.25 W/m^2 which corresponds to a maximum exposure time to solar UVR of 120 seconds or 2 minutes. Similarly, a UV index of 2 will have an effective solar UVR exposure of 0.05 W/m^2 which corresponds to a maximum exposure time to solar UVR of 600 seconds or 10 minutes.

The sensitivity of maximum exposure time to reach the standard UVR exposure energy limit of 30 Joules/m^2 to effective UVR exposure versus effective UVR exposure is presented in Figure 8-2. Absolute value of this sensitivity behaves as $30 / E^2_{effective}$ and approaches zero fast as effective UVR exposure increases.

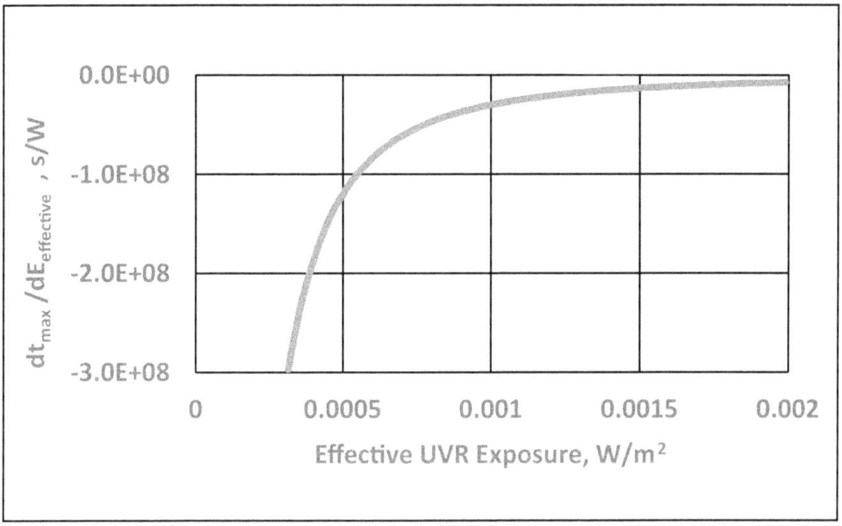

Figure 8-2 Sensitivity of maximum exposure time to reach the standard limit of 30 Joules/m^2 to effective UVR exposure versus effective UVR exposure for lamps and lighting equipment

The standard incandescent lamp produces white light from a tungsten filament heated by an electric current. For a 60-Watt incandescent lamp, the tungsten filament can reach a temperature of 2,773 K. In Figure 8-3 spectral blackbody emissive power of our Sun, i.e. *Sun's emissivity* = 1.0, is compared to the spectral graybody emissive power of a 60 Watt incandescent lamp with a tungsten filament, i.e. *tungsten's emissivity* = 0.42. At the source, our Sun's maximum radiation emission power is two orders of magnitude higher than the tungsten filament's. For our Sun, the maximum radiation emission power occurs at the wavelength of about 0.5 μm. For the 60 Watt incandescent lamp's tungsten filament, the maximum radiation emission power occurs at the wavelength of about 1.0 μm.

The geometry of human exposure to UVR emissions from a lamp source is very important. Above standard is applied to UVR exposures directed perpendicular to human body surfaces, i.e. skin or eyes. At the source of a 60-Watt incandescent lamp, about 0.1% of the spectral graybody emissive power is in the UVR wavelength range, i.e. 100 to 400 nm. UVR emission measurements are mostly made on a standard 60-Watt incandescent lamp at a distance of 20 cm. Using the erythemal weighing function, the 60-Watt incandescent lamp showed only 25×10^{-5} W/m^2 effective UVR exposure at 20 cm distance, see Ref. [17]. Also, nineteen different compact fluorescent lamps were measured for UVR emissions, see Ref. [17]. Using the erythemal weighing function, nineteen different compact fluorescent lamps had effective UVR exposure range of 0.00001 W/m^2 to 0.00046 W/m^2.

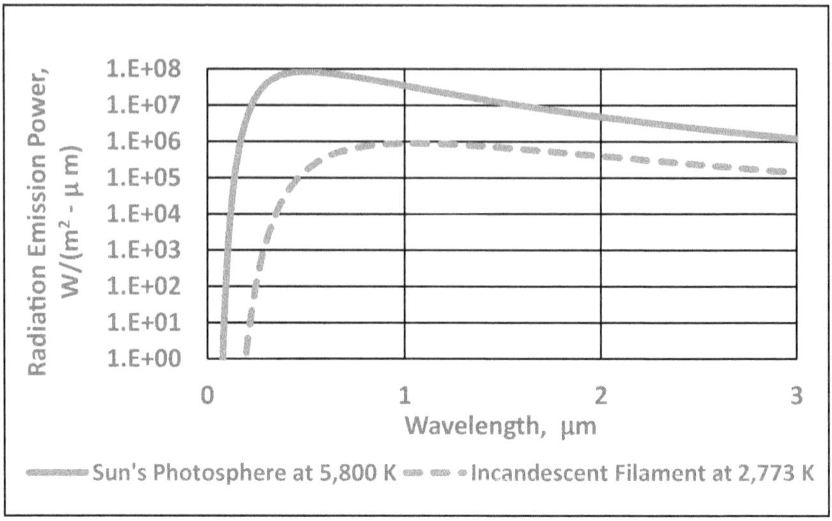

Figure 8-3 Spectral blackbody emissive power of our Sun is compared to the spectral graybody emissive power of a 60 Watt incandescent lamp with a tungsten filament

Since the safety standard for lamps and lighting equipment specify that the effective UVR exposure, i.e. spectrally weighted, has to be below 0.001 W/m^2, tests for incandescent and fluorescent lamps for general lighting show 2 to 100 fold less effective UVR exposure than the safety standard at 20 cm distance.

Incandescent and fluorescent lamps for general lighting do not cause effective UVR exposure problems as shown by very low values obtained from experiments. However, if there is no cover over the lamp and the glass envelope of the lamp is defective and has metal impurities in it, UVR emissions from defective lamps can cause significant cumulative damage to human eyes and skin.

Fluorescent sun lamps and sun beds emit mostly UVA and/or UVB radiation. They are popular for cosmetic tanning. They are characterized by high irradiance power in the UVA and UVB wavelength region, i.e. UVA and/or UVB > 0.15 W/m^2, which is more than 150 fold as compared to standard for incandescent and fluorescent lamps for general lighting, see Ref. [11]. Also effective UVR exposure, i.e. UVA plus UVB, should not exceed the standard value of 0.30 W/m^2. Exposure to fluorescent sun lamps and sun beds have cumulative damages to humans and can cause skin cancer, premature skin aging and severe damage to unprotected eyes.

Skin types are classified based on their susceptibility to their ability to tan as shown in Table 8-1. See Ref. [27]. Table 8-1 categorizes skin types I and II as easily burning and peeling skin response to UVR which are common in humans with fair and unexposed white skin. Table 8-1 categorizes skin types III and IV as moderately or minimally burning skin response to UVR which are common in humans with unexposed white or light brown skin. Table 8-1 categorizes skin

Table 8-1 Human skin response to UVR standard erythema dose (SED) values

Skin Type	UVR Sensitivity	UVR Burn Susceptibility	Skin Tanning
I	Very Sensitive	Always Susceptible: <2 SED	No Tan
II	Moderately Sensitive	High Susceptibility: 2 - 3 SED	Light Tan
III	Moderately Insensitive	Moderate Susceptibility: 3 - 5 SED	Medium Tan
IV	Insensitive	Low Susceptibility: 5 - 7 SED	Dark Tan
V	Insensitive	Very Low Susceptibility: 7 - 10 SED	Natural Brown Skin
VI	Insensitive	Extremely Low Susceptibility: SED > 10	Natural Black Skin

types V and VI as rarely or never burning skin response to UVR which are common in humans with brown or black skin.

UVR induced erythema on human skin is characterized in literature by standard erythema dose (SED) unit. One SED is defined as 100 J/m^2. See Ref. [27].

A lot of effective UV radiation exposure measurements of fluorescent sun tanning beds can be found in literature. Effective UV radiation exposure measurements were made over 400 fluorescent sun tanning beds in England in 2012. See Ref. [26]. Results of these erythema action spectrum weighted UVA plus UVB measurements, $E_{effective}$, ranged from 0.10 W/m^2 to 1.32 W/m^2. UVC is blocked out and it has no contribution in tanning devices. A very sensitive person to UVR with SED < 2 requirement, see Table 8-1, should not stay in the highest effective UVR exposure environment of a sun tanning bed more than 2.5 minutes and should not stay in the lowest effective UVR exposure environment more than 33 minutes.

Maximum exposure times versus effective UVR exposures for different SED's that are defined in Table 8-1 are shown in Figure 8-4. Maximum exposure times are obtained by using Equation 8-1 and by replacing 30 W/m^2 with the appropriate SED value in J/m^2. Most fluorescent sun tanning beds have maximum exposure times less than 0.5 hour. Again, it is a reminder that these maximum exposure times affect cumulatively to human health. Finally, now there are laws in many countries restricting usage of fluorescent sun tanning beds for people under the age of 18.

Sensitivity of maximum exposure time to reach the SED limits given in Table 8-1 to effective UVR exposure versus effective UVR exposure is shown in Figure 8-5. Absolute value of this sensitivity behaves as $SED / E^2_{effective}$ and approaches zero fast as effective UVR exposure increases.

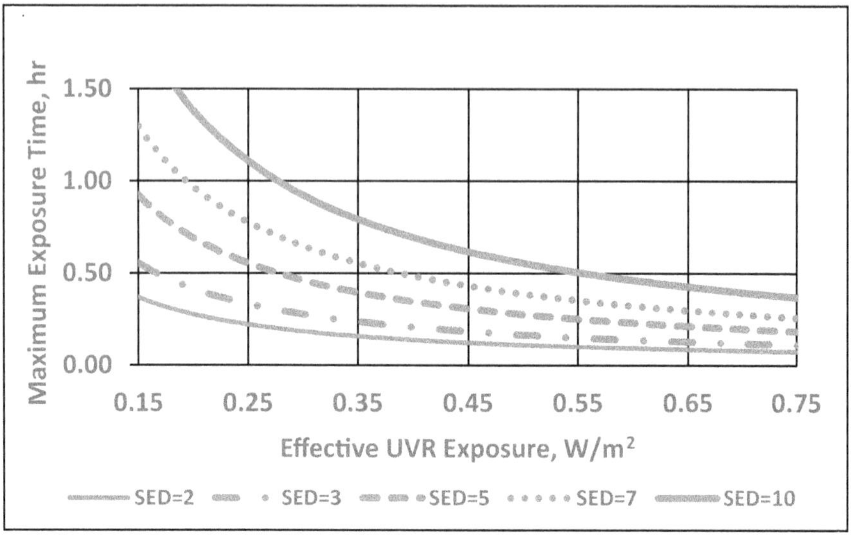

Figure 8-4 Maximum exposure times versus effective UVR exposures for different SED's

World Health Organization warns people to use extreme protection when the solar UV index exceeds 11 or $E_{effective}$ exceeds 0.275 W/m^2. On the other hand, most of the time, fluorescent sun tanning beds effective UVR exposures exceed the extreme solar UV index limit.

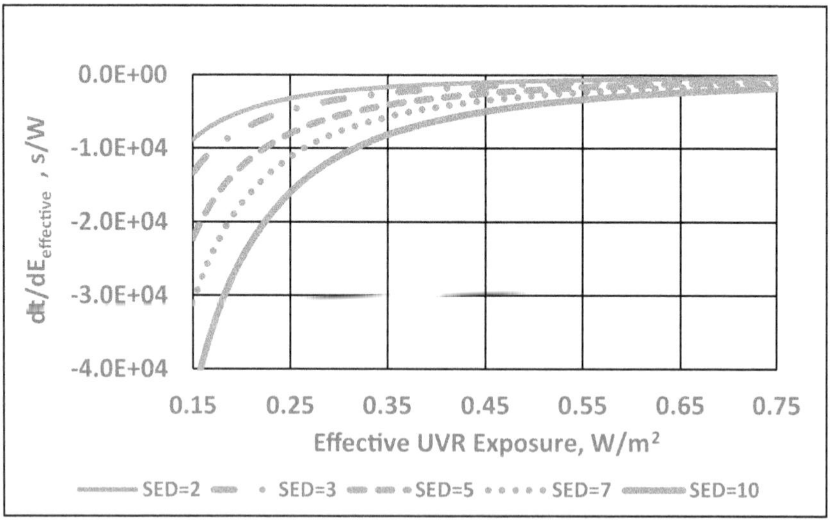

Figure 8-5 Sensitivity of maximum exposure time to reach the SED limit to effective UVR exposure versus effective UVR exposure

9

Radiation Heat Transfer from a Thermos Bottle

Heat transfer in a thermos bottle mainly involves radiation and to a much lesser extent conduction and convection. Dewar invented the thermos bottle at the beginning of the twentieth century which is simply a stainless steel inner bottle covered by another stainless steel outer bottle which are separated by vacuum. The outer wall of the inner bottle and the inner wall of the outer bottle are coated with silver in order to minimize radiation heat transfer between the fluid inside the thermos bottle and the outside environment. See Figure 9-1. Two silver coated walls facing each other reflect electromagnetic waves between themselves in the infrared region and minimize radiation heat transfer from the medium inside the inner bottle to the outside environment or vice versa. There is some heat loss, i.e. less than 2.58% of total heat loss in the present example, due to conduction and convection from the lid of the thermos bottle where a small portion of the inner surface of the stainless steel inner bottle contacts the lid.

Let us investigate heat transfer from a 0.5 liter capacity thermos bottle where the inner diameter of the inner bottle is 5.64 cm and the outer diameter of the outer bottle is 6.00 cm. The fluid height in the inner bottle is 20 cm. Stainless steel wall thickness of both the inner and the outer bottle is 0.05 cm. The gap between the outer wall of the inner bottle and the inner wall of the outer bottle is 0.08 cm neglecting the sputtered silver coating thicknesses on both surfaces.

The lid diameter of the thermos bottle is 4 cm and the lid height is 3 cm. The effective contact height between the inner surface of the stainless steel inner bottle and the lid is 0.5 cm.

Let us put very hot coffee, i.e. initially at T_{in} = 90 C or 363 K, into the above thermos bottle and see how the coffee cools down when the outside environment temperature is at T_{out} 20 C or 293 K. Assuming that the temperature of the coffee inside the thermos bottle is uniform at any instant of time, energy balance between the coffee and the outside environment can be expressed as shown in Equation 9-1.

$$\rho \times c_p \times V \times \frac{dT_{in}}{dt} = -\frac{A_{ss} \times (T_{in} - T_{out})}{\sum_i R_i} - \frac{A_{lid} \times (T_{in} - T_{out})}{\sum_j R_j} \qquad 9\text{-}1$$

where ρ is the density of the coffee in kg/m^3, c_p is the specific heat of coffee at constant pressure in J/(kg-K), V is the volume of coffee in m^3, T_{in} is the temperature of the coffee in K at a time t which is in seconds. A_{ss} is stainless steel outer bottle's surface area in m^2. T_{out} is the temperature of outside environment in K. $\sum_i R_i$ is the sum of radiation, conduction and natural convection heat transfer resistances through stainless steel bodies of the thermos bottle between coffee and the outside environment in W/(m^2 - K). A_{lid} is stainless steel bottle's surface area that contacts the insulating lid in m^2. $\sum_j R_j$ is the sum of conduction heat transfer resistance from the stainless steel inner bottle's inner surface contact area to the insulating lid's top surface and the natural convection heat transfer resistance from top of the lid's surface to outside environment in W/(m^2 – K). Radiation heat transfer from the top surface of the coffee to the lid's bottom surface is neglected assuming very low infrared radiation absorptivity at the bottom of the lid.

In order to evaluate infrared radiation heat transfer between two inner surfaces of two stainless steel bottles, it is assumed that both surfaces are gray and opaque, namely the sum of emissivity and reflectivity are one for both inner surfaces and there is no transmissivity. Irradiation of gray and opaque two large parallel surfaces are given in Equation 9-2. See Ref. [33] for derivation of Equation 9-2.

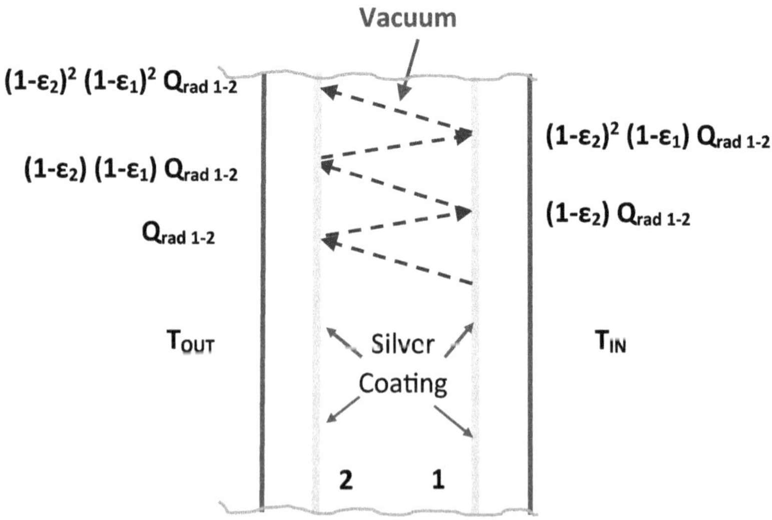

Figure 9-1 Cross-sectional view of stainless steel stainless inner bottle wall and stainless steel outer bottle wall. The outer wall of the inner bottle (surface 1) and the inner wall of the outer bottle (surface 2) are coated with silver.

$$Q_{1-2} = A_{ss} \times \sigma \times \frac{\left(T_{in}^4 - T_{out}^4\right)}{\frac{1}{\varepsilon_1} + \frac{1}{\varepsilon_2} - 1} \qquad 9\text{-}2$$

where "1" represents the outer silver coated surface of the inner bottle and "2" represents the inner silver coated surface of the outer bottle. ε_1 and ε_2 are emissivities of two silver coated surfaces. In the present analysis, it is always assumed that $\varepsilon_1 = \varepsilon_2$ and they always have a constant value which is determined at the average of the initial coffee temperature and the outside environment temperature. σ is the Stefan-Boltzmann constant which is 5.67×10^{-8} W/(m^2 - K^4).

Now let us compare resistances to heat flow from the coffee to outside environment through stainless steel inner and outer bottles and eliminate the ones that have negligible contribution to heat flow. Radiation heat transfer resistance is obtained by rewriting Equation 9-2 as shown in Equation 9-3.

$$R_{radiation\ 1\text{-}2} = \frac{\left(T_{in} - T_{out}\right) \times \left(\frac{1}{\varepsilon_1} + \frac{1}{\varepsilon_2} - 1\right)}{\sigma \times \left(T_{in}^4 - T_{out}^4\right)} \qquad 9\text{-}3$$

For $\varepsilon_1 = \varepsilon_2 = 0.01$ which is about the lowest possible emissivity that can be obtained for a silver coated stainless steel surface by sputtered deposition, for $T_{in} = 363\ K$, and for $T_{out} = 293\ K$, using Equation 9-3, radiation heat transfer resistance between two silver coated stainless surfaces is given in Equation 9-4.

$$R_{\ radiation\ 1-2} = 24.59\ W / \left(m^2 - K\right) \qquad 9\text{-}4$$

For conduction heat transfer resistance through walls of stainless steel inner bottle and through walls of stainless steel outer bottle, let us assume that wall thicknesses are the same and they are 0.01 cm thick. Also stainless steel thermal conductivities are the same and it is 15 W/(m - K). Conduction heat transfer resistance through the walls of both bottles are given in Equation 9-5.

$$R_{conduction\ ss\ walls} = \frac{L_{ss}}{k_{ss}} = 6.67 \times 10^{-6} \left(m^2 - K\right) / W \qquad 9\text{-}5$$

Average natural convection heat transfer between the coffee and the vertical wall surface of the inner bottle is in the laminar flow region, i.e. Grashof number $< 10^9$, as shown in Equation 9-6. See Ref. 13, Chapter 9 for details.

$$R_{natural\ conv.coffee\ to\ ss} = \frac{1}{h_{coffee\ to\ ss}} = 5.5 \times 10^{-4} \left(m^2 - K\right) / W \qquad 9\text{-}6$$

Similarly, average natural convection heat transfer between the outside air and the vertical wall surface of the outer bottle is in the laminar flow region, i.e. Grashof number < 10^9, as shown in Equation 9-7. See Ref. 13, Chapter 9 for details.

$$R_{natural\ conv.\ ss\ to\ air} = \frac{1}{h_{air\ to\ sss}} = 9.5\times10^{-2}\left(m^2 - K\right)/W \qquad 9\text{-}7$$

If different modes of heat transfer resistance values are compared in Equations 9-4 through 9-7, it can be deduced that the radiation heat transfer resistance value is several orders of magnitude higher than the others. So, in the present thermos bottle heat transfer analysis from the coffee to the outside environment through the stainless steel walls, only radiation heat transfer mode will be considered.

For the conduction heat transfer resistance from the stainless steel inner bottle's inner surface contact area to the insulating lid's top surface, let us assume that lid's characteristic length is 0.07 cm. Also the lid is made out of an insulating material like cork and its thermal conductivity is 0.02 W/(m – K). Conduction heat transfer resistance through the lid is given in Equation 9-8.

$$R_{conduction\ from\ lid} = \frac{L_{lid}}{k_{lid}} = 3.50\left(m^2 - K\right)/W \qquad 9\text{-}8$$

Average natural convection heat transfer between the outside air and top of lid's surface is again in the laminar flow region, i.e. Grashof number < 10^9, and its resistance has a lesser contribution than the one given in Equation 9-7, i.e. 0.064 (m^2-K)/W). Since the average natural convection heat transfer resistance through the lid is 1.8% of the total heat transfer resistance through the lid, it will be included into the T_{in} versus time calculations.

Temperature of coffee versus time is determined by finite difference method using one hour time intervals. The finite difference method is also exercised using one minute time intervals. One minute time intervals resulted in 0.2% higher temperatures after 24 hours as compared to one hour time intervals. So one hour time interval is used in calculating temperature of coffee versus time using Equation 9-9.

$$\rho\times c_p\times V\times\frac{d\left(T_{in@k\text{-}1} - T_{in@k\text{-}1}\right)}{d\left(t_k - t_{k-1}\right)} = -\frac{A_{ss}\times\left(T_{in@k\text{-}1-T_{out}}\right)}{R_{radiation\ 1\text{-}2\ @k\text{-}1}} - \frac{A_{lid}\times\left(T_{in@k\text{-}1-T_{out}}\right)}{R_{conduction\ from\ lid\ @k\text{-}1} + R_{convection\ lid\ top\ to\ air\ @k\text{-}1}} \qquad 9\text{-}9$$

Temperature of coffee versus time is calculated using Equation 9-9 and an internal energy of coffee per unit temperature of $\rho\times c_p\times V = 1970.71$ *J/K*. Temperature drop of coffee versus time (for 3 days) are shown in Figure 9-2 for different silver coating emissivity values and also for stainless steel surfaces 1 and 2 without any silver coating at an emissivity of 0.22. As the stainless steel's surface emissivity increases, temperature drop of coffee with time also increases as expected. When the silver coating emissivity is 0.01, the temperature drop of coffee in three days is only 15 C. On the other hand, if there is no silver coating on outer wall of inner bottle and inner wall of outer bottle, temperature of coffee drops down to the outside environment temperature of 20 C in three days.

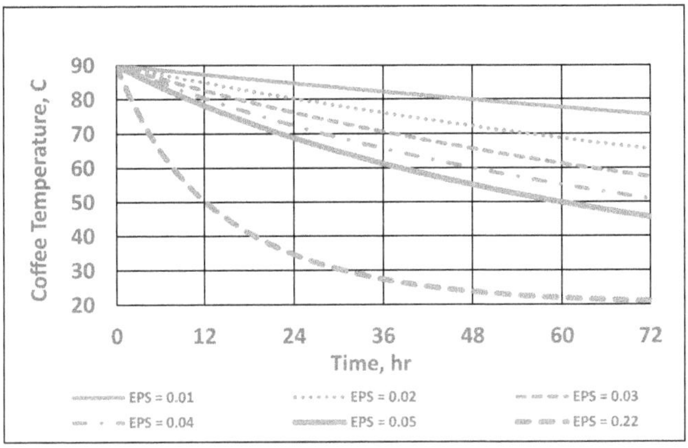

Figure 9-2 Temperature of coffee versus time for different silver coating emissivities and for stainless steel surfaces 1 and 2 without any silver coating.

In the present analysis, the heat loss from lid of the thermos bottle accounted on the average only 2.58% of the total heat loss. On the average, heat loss due to thermal radiation from stainless steel walls of the thermos bottle was 97.42%.

Sensitivities of coffee temperature versus silver coating emissivities are shown in Figure 9-3 at 24 hours and at 72 hours. Absolute values of this sensitivity increases as time increases. In 24 hours, as silver coating emissivity increases from 0.015 to 0.05, the absolute value of sensitivity decreases by 23%. In 72 hours, as silver coating emissivity increases from 0.015 to 0.05, the absolute value of sensitivity decreases by 53%.

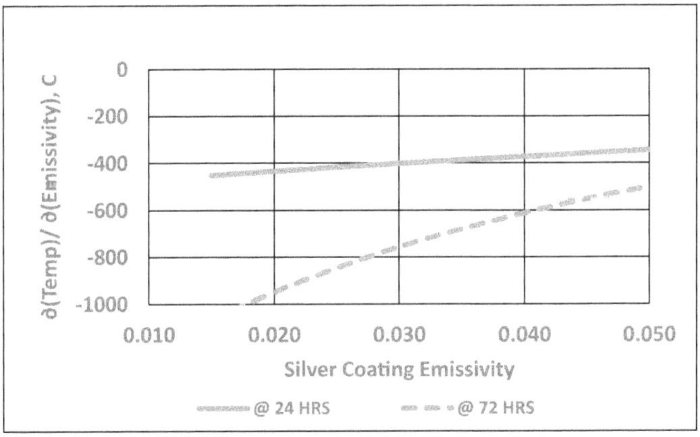

Figure 9-3 Sensitivities of coffee temperature to silver coating emissivity versus silver coating emissivity at 24 hours and at 72 hours.

Average time constant over 72 hours versus emissivity of outer wall of inner bottle and inner wall of outer bottle is shown in Figure 9-4. At each time interval of one hour, the time constant was calculated using Equation 9-10.

$$Time\ Constant = \frac{\rho \times c_p \times V}{h_{radiation\ 1\text{-}2} \times A_{ss} + \left(\frac{k_{lid}}{L_{lid}} + h_{convection\ lid\ top\ to\ air} \right) \times A_{lid}} \quad 9\text{-}10$$

Average time constant represents the time that the coffee temperature drops 37% from its initial value of 90 C. Since the radiation heat transfer coefficient between surfaces 1 and 2 are time dependent in Equation 9-10, a time averaged time constant is used with minimum error to determine the effect of emissivity variation to the coffee temperature drop. The standard deviation of instantaneous time constants varied from 0.26 hour at an emissivity of 0.01 to 1.28 hours at an emissivity of 0.05.

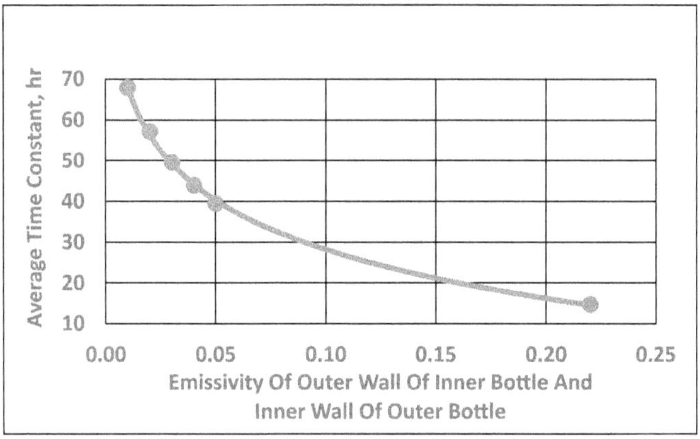

Figure 9-4 Average time constant versus emissivity of outer wall of inner bottle and inner wall of outer bottle.

10

Radiation and Laminar Forced Convection Heat Transfer During a Vehicle's Paint Curing Process

Painting a vehicle involves several layers such as electrocoating that is applied to the virgin metal body, primer, base coat, clear coat, sealants and structural adhesives. In order to achieve the best paint performance, each layer of paint has to be cured in a bake oven according to paint supplier's bake temperature versus time specifications. Accurate control of bake temperature versus time specifications is critical to the performance of a coating both physically and cosmetically. During the baking process, complete cross linking of the paint has to be achieved. In the present example, let us investigate a paint curing bake oven where heat is supplied to a particular vehicle's painted body by hot air flow, i.e. by laminar forced convection heat transfer and by infrared radiation heat transfer lamps.

Initially bake oven temperature mapping is done on the surface of a similar test vehicle, by using thermocouples on different sections of the test vehicle, for the desired bake temperature versus time characteristics for a certain layer of paint. In the present example, infrared radiation heat lamps will have a fixed power setting. Air flowing over the body of the vehicle in the laminar forced convection region at constant temperature will control painted surface's bake temperature versus time characteristics. Bake oven's walls are also kept at a constant temperature. Once the temperature versus time mapping of the test vehicle is satisfactory per the paint supplier's bake temperature versus time specifications and transient temperature conditions have passed and steady-state temperature versus time mapping is achieved, the test vehicle can be replaced by the freshly painted vehicle in the bake oven. For a certain layer of fresh paint on a vehicle similar to the test vehicle has to be cured at a specified temperature for a specified time.

For example, for the primer coating layer of the vehicle, the paint supplier specifies curing at 210 $C \pm 15\ C\ or$ (483 $K \pm 15K$) for 30 minutes. Under steady-state conditions, conservation of energy on the painted surface is given in Equation 10-1.

$$\alpha \times Q_{IR-in} = \varepsilon \times \sigma \times \left(T_{surf}^4 - T_{oven}^4\right) + h_{conv} \times \left(T_{surf} - T_{air}\right) \qquad 10\text{-}1$$

Equation 10-1 assumes that the painted surface's local temperature, T_{surf}, is a constant in time and does not vary through previously painted layers and through the metal body. α is the absorptance of the painted surface for the infrared radiation energy incoming from heat lamps and its value is 0.85. $Q_{IR\text{-}in}$ is the infrared radiation power incoming from heat lamps at a constant value of 3,000 W/m^2. ε is the emissivity of the painted surface at 0.85. T_{oven} is the bake oven's constant wall temperature at 30 C or 303 K. T_{air} is the constant free stream temperature of air flowing over the painted surface at 25 C or 298 K.

h_{conv} *in W/(m^2 – K)* is the local forced convection heat transfer coefficient in the laminar flow region, i.e., $Re_x = \frac{U \times X}{v} < 5 \times 10^5$, given in Equation 10-2. In the present analysis, the vehicle's painted surface is assumed to be a flat plate. Re_x is the local Reynolds number at distance X in meters from the leading edge of the vehicle. U is the air flow's free stream velocity in meters per second. v is the kinematic viscosity of air in m^2/s at film temperature, namely at $0.5 \times (T_{surf} + T_{air})$.

$$h_{conv} = \left(\frac{k}{x}\right) \times 0.332 \times Re_x^{0.5} \times Pr^{0.333} \qquad 10\text{-}2$$

where k is the thermal conductivity of air flowing over the painted surface in W/(m-K) and $Pr = \frac{\mu \times c_p}{k}$ is the dimensionless Prandtl number of air flowing over the painted surface. μ is the viscosity of air flowing over the painted surface in (N-s)/m^2 and c_p is the specific heat of air at constant pressure flowing over the painted surface in J/(kg-K). In the present analysis all air properties are taken at film temperature.

The average value of the local heat transfer coefficient over a vehicle length L can be obtained by integrating Equation 10-2 from $X = 0$ to $X = L$. The average value of heat transfer coefficient, $\bar{h}_{conv}$, over a vehicle length L in the laminar flow region is given in Equation 10-3. In the laminar flow region, the average value of heat transfer coefficient, $\bar{h}_{conv}$, over a vehicle length L is twice the local heat transfer coefficient value at length L.

$$\bar{h}_{conv} = 2 \times h_{conv@x=L} \tag{10-3}$$

Average value of forced convection heat transfer coefficient, $\bar{h}_{conv}$, over a vehicle length L in the laminar flow region is shown in Figure 10-1. $\bar{h}_{conv}$ versus U values are obtained from Equations 10-2 and 10-3 for a vehicle with a length of 4.42 meters. Vehicle length averaged forced convection heat transfer coefficient in the laminar flow region at the upper limit, i.e. $Re_L = 5 \text{ X } 10^5$, has a value of 3.10 W/(m^2-K) with a limiting air flow free stream velocity of 2.83 m/s. The vehicle's length averaged forced convection heat transfer coefficient increases as $U^{0.5}$ with increasing free stream air velocity.

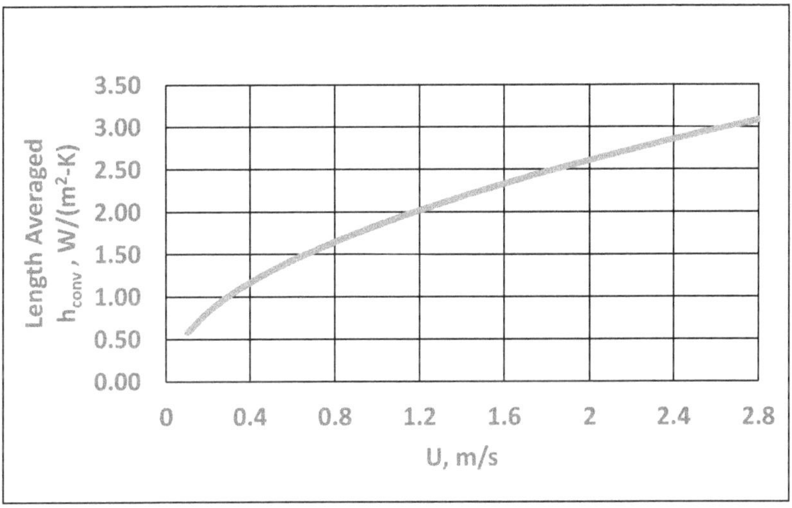

Figure 10-1 Average value of forced convection heat transfer coefficient, $\bar{h}_{conv}$, over a vehicle's length L versus free stream air velocity in the laminar flow region.

Once the vehicle's length averaged forced convection heat transfer coefficient is known, Equation 10-1 can be rewritten as follows, shown in Equation 10-4, in order to calculate for T_{surf} by iteration.

$$f\left(T_{surf}\right) = -\alpha \times Q_{IR-in} + \varepsilon \times \sigma \times \left(T_{surf}^4 - T_{oven}^4\right) + \bar{h}_{conv} \times \left(T_{surf} - T_{air}\right) = 0 \tag{10-4}$$

Using Equation 10-4, a typical iteration result is shown in Figure 10-2 for $U = 0.60$ *m/s* and for $\bar{h}_{conv} = 1.43$ $W/(m^2 - K)$ where at $f(T_{surf}) = 0$ crossing $T_{surf} = 485.95$ *K*. In the present analyses, local painted surface temperatures are obtained using local heat transfer coefficients, $\bar{h}_{conv}$, instead of vehicle's length averaged heat transfer coefficients, $\bar{h}_{conv}$, in Equation 10-4.

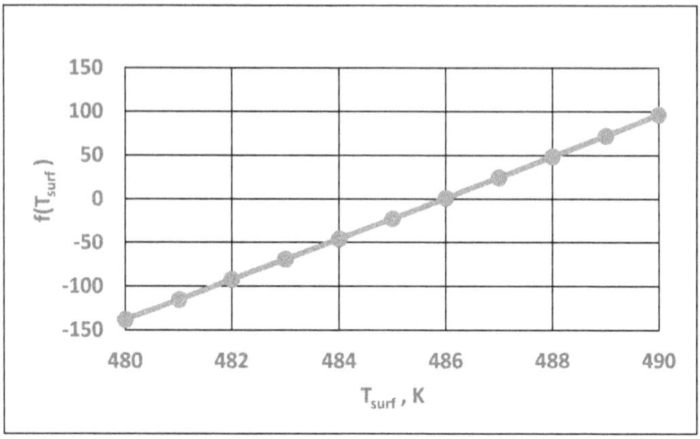

Figure 10-2 $f(T_{surf})$ versus T_{surf} at $U = 0.60\ m/s$ and for $\overline{h}_{conv} = 1.43\ W/(m^2 - K)$ with $f(T_{surf}) = 0$ crossing at $T_{surf} = 485.95\ K$.

For example, for curing the primer coating layer of the vehicle, the paint supplier specified temperature is $210\ C \pm 15\ C\ or\ (483\ K \pm 15K)$ for 30 minutes. Under steady-state conditions and with 3,000 W/m^2 infrared radiation power input from heat lamps, average painted surface bake temperature versus free stream air velocity over the vehicle is shown in Figure 10-3. These results are obtained by using vehicle's length averaged forced convection heat transfer coefficients from Figure 10-1. Examining Figure 10-3 shows that, if the free stream air velocity over the vehicle is set at 1.0 m/s, the vehicle's length averaged painted surface bake temperature of 483 K should be achieved. This average painted surface bake temperature is the nominal bake temperature that the paint supplier specified for curing this particular primer coat layer.

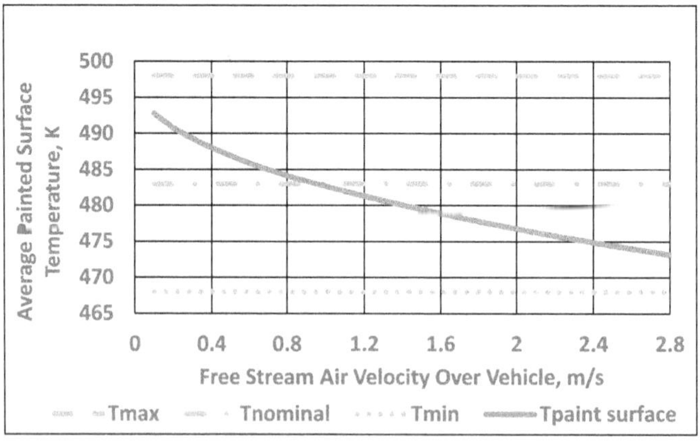

Figure 10-3 For primer coating layer, average painted surface temperature versus free stream air velocity over the vehicle using vehicle's length averaged laminar forced convection heat transfer coefficients from Figure 10-1.

However, bake oven temperature mapping that was done on the surface of a similar test vehicle, showed that painted surface temperatures at the front section of the vehicle will be below the lower specification limit of 468 K for U=1.0 m/s. So instead of using the vehicle's length averaged value of forced convection heat transfer coefficient, the local forced convection heat transfer coefficient in the laminar flow region, h_{conv}, should be used in the present calculations. Local h_{conv} is shown in Figure 10-4 as a function of length along the vehicle. Local h_{conv} is calculated using Equation 10-2 for U=1.0 m/s. Also, in Figure 10-4, local h_{conv} is also compared to the vehicle's length averaged forced convection heat transfer coefficient. Figure 10-4 shows that the local heat transfer coefficient is much higher than the vehicle's length averaged heat transfer coefficient at the front section of the vehicle which causes painted surface temperatures at the front section of the vehicle to be below the lower specification limit of 468 K.

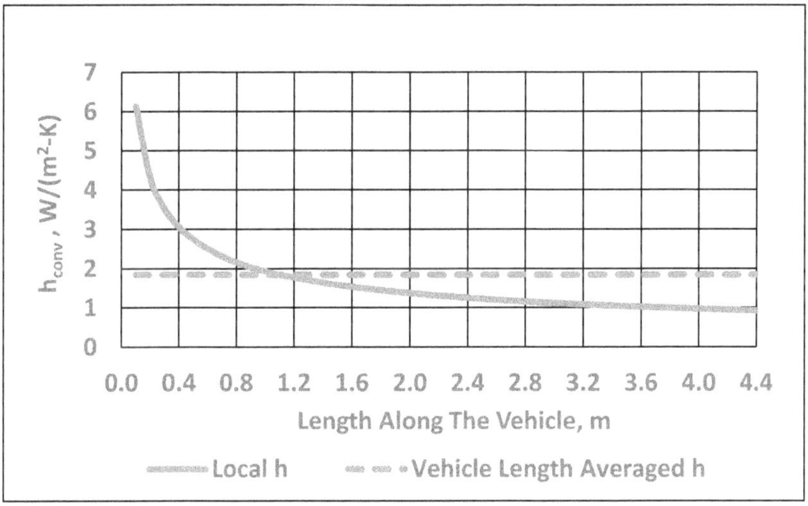

Figure 10-4 Local h_{conv} and the vehicle's length averaged laminar forced convection heat transfer coefficient for U=1.0 m/s.

Painted surface temperatures for primer coating layer, obtained using local heat transfer coefficients and the vehicle's length averaged heat transfer coefficient for U=1.0 m/s from Figure 10-4, are shown in Figure 10 5. The front section of the vehicle, i.e. first 25 cm, will have temperatures below the lower specification limit, if the air flow free stream velocity is set at U=1.0 m/s. Using thermocouples placed at the front section of the test vehicle, these undesirable bake temperature conditions should be verified for U=1.0 m/s.

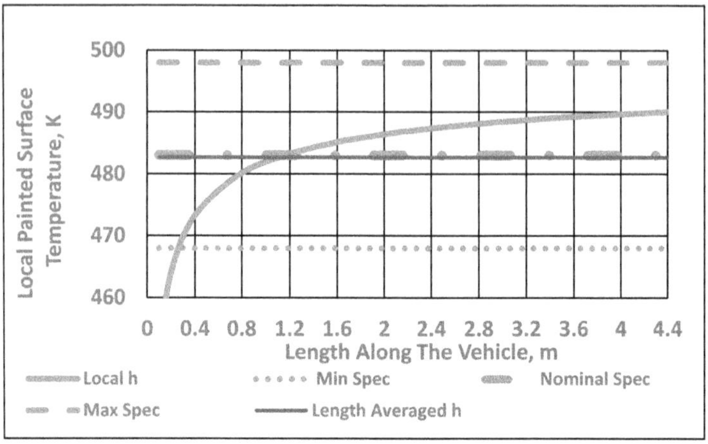

Figure 10-5 Painted surface temperatures for the primer coat layer that are obtained using local heat transfer coefficients and the vehicle's length averaged heat transfer coefficient for U=1.0 m/s.

Then, local temperature calculations at X = 0.1 m are used to determine the appropriate free stream air velocity so that during the curing process local temperatures at the front of the vehicle do not fall below the minimum cure temperature specification. Local temperatures at X = 0.1 m obtained by using local heat transfer coefficient versus free stream air velocity are shown in Figure 10-6. Also the local temperature at X = 0.1 m is very sensitive to changes in free stream air velocity as shown in Figure 10-7. Absolute value of the sensitivity to free stream air velocity increases two fold when U decreases from 0.6 m/s to 0.2 m/s.

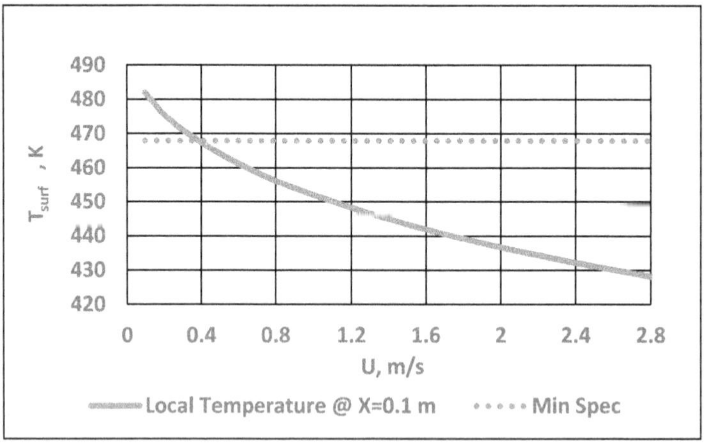

Figure 10-6 For primer coat layer curing, local temperature at X = 0.1 m obtained by using local heat transfer coefficient versus free stream air velocity.

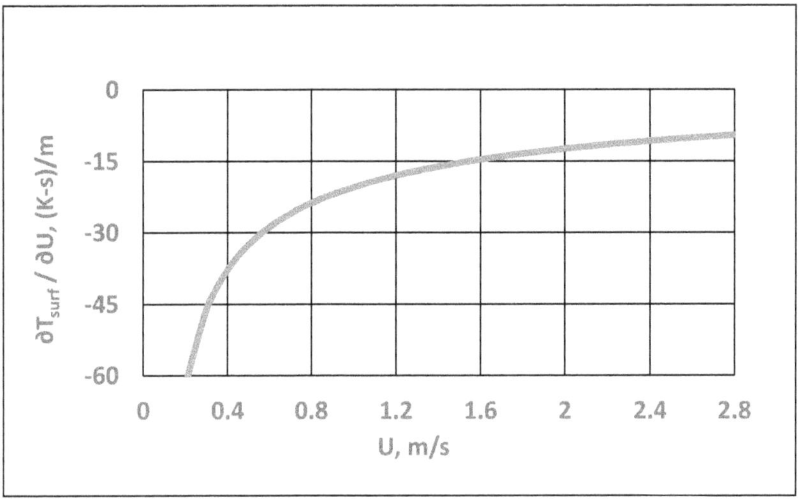

Figure 10-7 For primer coat layer curing, sensitivity of local temperature at X = 0.1 m to free stream air velocity versus free stream air velocity.

In order to correct for the out of specification painted surface temperatures in the front of the vehicle for primer coat layer curing, a free stream air velocity of 0.38 m/s is chosen from Figure 10-6. The bake oven temperature mapping is redone using the free stream air velocity of 0.38 m/s. Local h_{conv} is shown in Figure 10-8 as a function of length along the vehicle for free stream air velocity of 0.38 m/s. Local h_{conv} is calculated using Equation 10-2. Also, in Figure 10-8, local h_{conv} is compared to the vehicle's length averaged value of forced convection heat transfer coefficient. Local heat transfer coefficients and the vehicle's length averaged heat transfer coefficient is 1.6 times lower at U=0.38 m/s than at U=1.0 m/s.

Painted surface temperatures are obtained using local heat transfer coefficients, namely h_{conv} instead of $\overline{h}_{conv}$, in Equation 10-4. Painted surface temperatures that are obtained for the primer coat layer using local heat transfer coefficients and the vehicle's length averaged heat transfer coefficient for U=0.38 m/s are shown in Figure 10-9. Under this free stream air velocity condition, the painted surface temperatures are within the specification limits all throughout the length of the vehicle.

Local painted surface temperature results for the primer coat layer that are shown in Figure 10-9 for free stream air velocity at 0.38 m/s are verified by performing temperature versus time mapping on the test vehicle using thermocouples at specified vehicle locations.

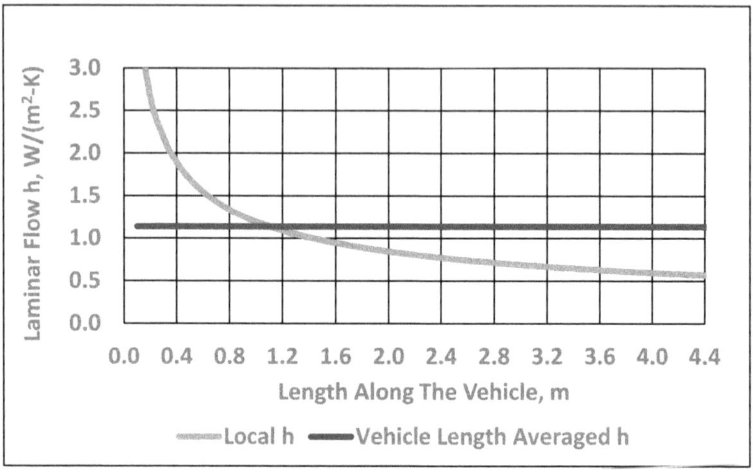

Figure 10-8 Local h_{conv} and the vehicle's length averaged laminar forced convection heat transfer coefficient for U=0.38 m/s.

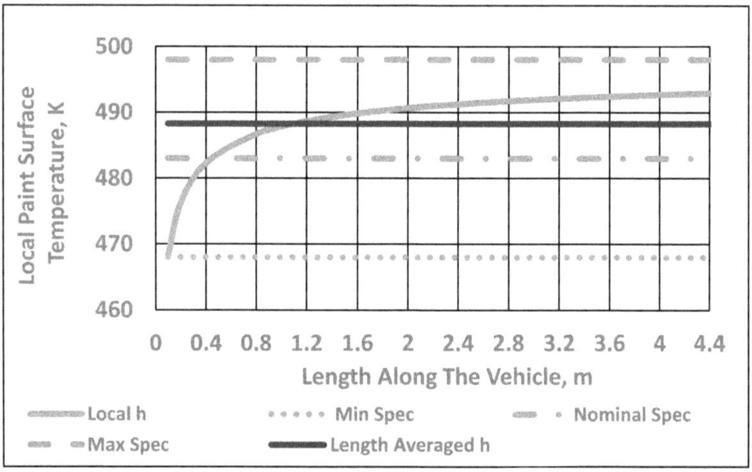

Figure 10-9 Painted surface temperatures for the primer coat layer that are obtained using local heat transfer coefficients and the vehicle's length averaged heat transfer coefficient for U=0.38 m/s.

Let us analyze another painted layer curing case for the same vehicle. The paint supplier for the base coat layer which is the cosmetic layer that provides the color of the finished product specifies a thermal cure specification of 137 $C \pm 20$ C *or* (410 $K \pm 20$ K) for 60 minutes.

Under steady-state conditions and with 1,500 W/m^2 infrared radiation power input from heat lamps, average painted surface bake temperature versus free stream air velocity over the vehicle are calculated using vehicle's length averaged laminar forced convection heat transfer coefficients and results are shown in

Figure 10-10. For the base coat curing specifications, we are inclined to choose a free stream air velocity of 2.4 m/s by examining Figure 10-10.

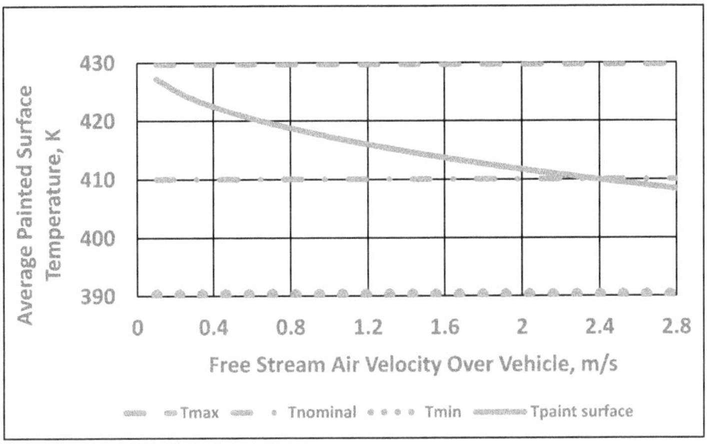

Figure 10-10 For base coat layer, average painted surface temperature versus free stream air velocity over the vehicle using vehicle's length averaged laminar forced convection heat transfer coefficients.

Choice of a 2.4 m/s free stream air velocity gives us local painted surface temperatures shown in Figure 10-11 for the base coat layer curing case. The base coat layer at the front section of the vehicle will again have temperatures below the lower specification limit, if the air flow free stream velocity is set at U=2.4 m/s as shown in Figure 10-11.

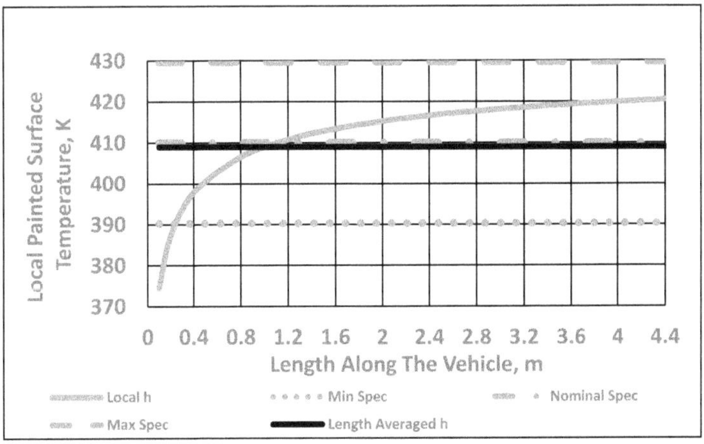

Figure 10-11 Painted surface temperatures for the base coat layer that are obtained using local heat transfer coefficients and the vehicle's length averaged heat transfer coefficient for U=2.4 m/s.

Again local temperature calculations at X = 0.1 m are used to determine the appropriate free stream air velocity so that during the curing process local temperatures at the front of the vehicle do not fall below the minimum cure temperature specification. Local temperatures at X = 0.1 m obtained by using local heat transfer coefficient versus free stream air velocity are shown in Figure 10-12. Local temperature at X = 0.1 m is very sensitive to changes in free stream air velocity as shown in Figure 10-13.

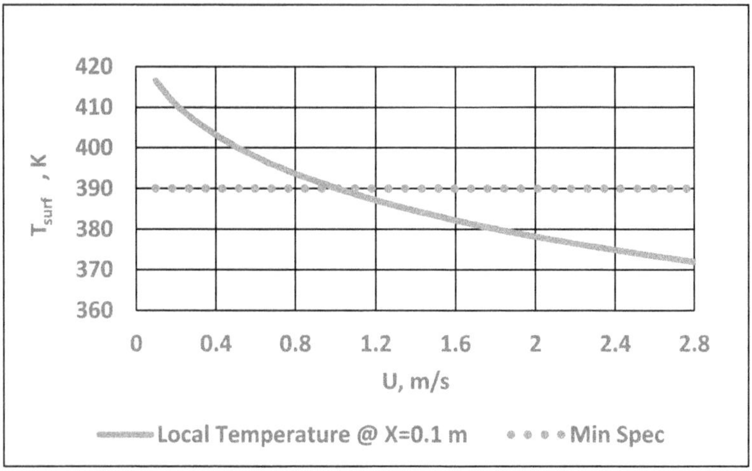

Figure 10-12 For base coat layer curing, local temperature at X = 0.1 m obtained by using local heat transfer coefficient versus free stream air velocity.

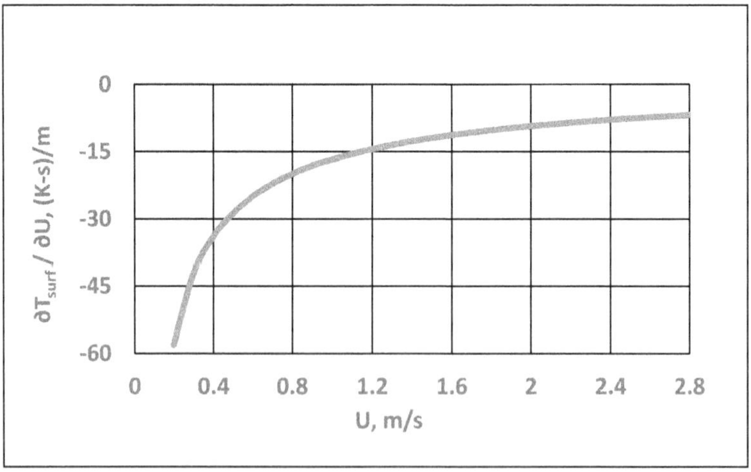

Figure 10-13 For base coat layer curing, sensitivity of local temperature at X = 0.1 m to free stream air velocity versus free stream air velocity.

Using Figure 10-12, a lower air flow free stream velocity of 0.7 m/s should bring the whole painted vehicle's surface temperatures within the base coat layer thermal cure specification of 137 $C \pm 20\ C\ or\ (410\ K \pm 20\ K)$. Base coat layer surface temperatures obtained using local heat transfer coefficients and the vehicle's length averaged heat transfer coefficient for U=0.7 m/s are shown in Figure 10-14. U=0.7 m/s air flow condition combined with a constant 1,500 W/m^2 infrared radiation power input, base coat layer surface temperatures are within the specification limits all throughout the length of the vehicle.

Local painted surface temperature results for the base coat layer that are shown in Figure 10-14 for free stream air velocity setting at 0.7 m/s are verified by performing temperature versus time mapping on the test vehicle using thermocouples at specified vehicle locations.

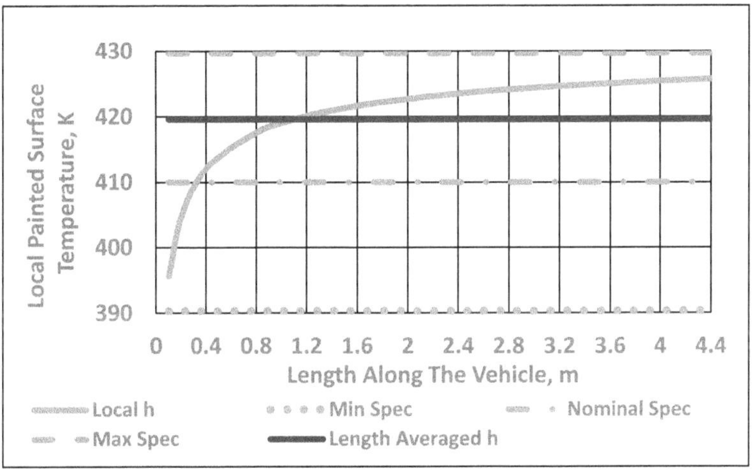

Figure 10-14 Painted surface temperatures for the base coat layer that are obtained using local heat transfer coefficients and the vehicle's length averaged heat transfer coefficient for U=0.7 m/s.

I would like to emphasize that these kinds of heat transfer calculations to control curing of painted surfaces of a vehicle in a bake oven will provide ballpark control values for the bake oven's operational variables, since a lot of assumptions are made during above calculations about bake oven conditions, air flow parameters, painted surface parameters and infrared radiation energy inputs from heat lamps. Final settings of a bake oven's operational variables should always be verified by performing temperature versus time mapping on a test vehicle using thermocouples at specified vehicle locations.

11

Sun's Irreplaceable Energy Source Optimization for a PV Collector Panel's Tilt Angle

In 2020, World energy consumption is estimated to be about 167,786 TeraWh/year. Only 1.25% of this energy consumption, namely 2,093 TeraWh/year, will come from our Sun through photovoltaic (PV) panels. Projection of yearly World energy consumption to 2040 show not much increase from the 2020 consumption. However, by 2040, energy coming from our Sun increases eight-fold to 9.66% of World energy consumption to 16,747 TeraWh/year mostly replacing coal and oil. This bright future in solar radiation energy consumption makes everyone eager to learn about PV panel system design and optimization of solar radiation energy conversion into electrical energy. This optimization process starts with the knowledge of solar radiation energy available at a particular location and at a particular time on Earth.

For example, a PV collector panel in San Diego, CA at the latitude of 32.73 degrees North, that is facing South at an optimum tilt angle from horizontal, will receive a total radiant energy, i.e. sum of direct beam, diffuse and ground reflected energy irradiation, of 2,091 kWh/(m^2-year). A similar PV collector panel facing South in Anchorage, AK at the latitude of 61.17 degrees North, at its optimum tilt angle, will only receive a total radiant energy of 1,086 kWh/(m^2-year).

Yearly electricity generation of a PV panel system, *E,* in kWh/year can be estimated by using Equation 11-1.

$$E = SR_{total} \times A \times PV_{yield} \times EFF \qquad 11\text{-}1$$

where SR_{total} is the total radiant energy that a South facing PV collector panel receives at an optimum tilt angle in a year in kWh/(m^2-year), A is the total surface area of PV collector panels in m^2, PV_{yield} is the PV collector panel yield and

EFF is the sum of efficiency losses in a PV collector panel system, namely in the inverter, wiring, connectors, temperature rise, shading, etc.

In this chapter, maximum available SR_{total} for South facing a PV collector panel will be determined from available solar radiation data by finding out the optimum and fixed tilt angle for San Diego, CA and for Anchorage, AK.

The PV collector panel yield has been improving continuously. PV_{yield} varies between 15% to 25% in today's PV technologies. PV_{yield} is specified by the manufacturer as the direct power output under standard test conditions divided by the panel's area. For example, if a PV panel with an area of 1.5 m^2 is tested under an irradiation power of 1,000 W/m^2, at an atmospheric air density of 1.5, i.e. AM1.5 (an irradiation zenith angle of 48.2 degrees), at 25 C temperature, at 1 m/s wind speed, and it produces 300 Watts of peak power, its yield is $\frac{0.300}{1.5} = 0.20\ or\ 20\%$. A value of $PV_{yield} = 0.20$ will be used in the present analysis.

EFF is the sum of PV panel system efficiencies which can vary from a minimum 50% to a maximum of 90%. A value of *EFF* = 0.70 will be used in the present analysis.

There are a lot of data available for solar radiation on Earth's surface on flat plates at different times and at different locations all over the World. In the present analysis, solar radiation data from Ref. [25] will be used.

Monthly averaged daily solar radiation data on a PV collector panel facing South at five different tilt angles in San Diego, CA are shown in Figure 11-1. Yearly averaged daily solar radiation data from Figure 11-1 are given in Figure 11-2. Best fit curve to data in Figure 11-2 shows that the maximum yearly solar radiation is achieved when a South facing PV collector panel is tilted 30.07 degrees from the horizontal position in San Diego, CA. Under this tilt angle setting, a PV collector panel will receive 5.68 kWh/(m^2-day) or 2,074 kWh/(m^2-year) of solar radiation.

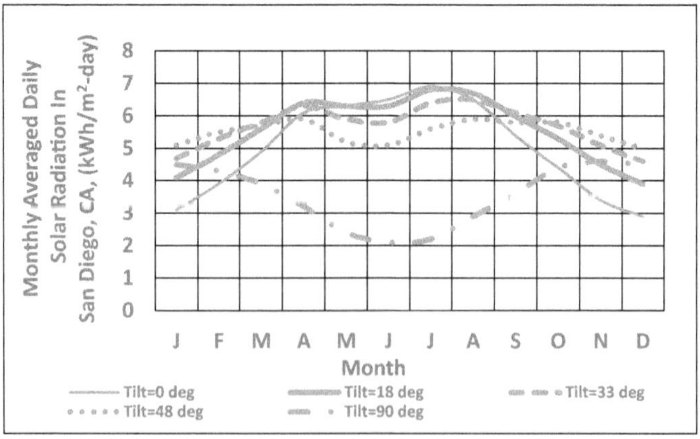

Figure 11-1 Monthly averaged daily solar radiation on a PV collector panel facing South at different tilt angles in San Diego, CA. at latitude of 32.73 degrees North, longitude of 117.17 degrees West, 9 meters elevation and 1014 millibars mean pressure.

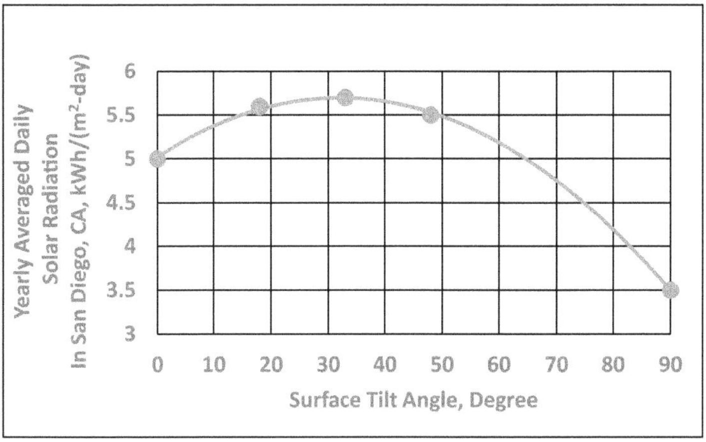

Figure 11-2 Yearly averaged daily solar radiation on a PV collector panel facing South at different tilt angles in San Diego, CA.

Monthly averaged daily solar radiation data on a PV collector panel facing South at five different tilt angles in Anchorage, AK are shown in Figure 11-3. Yearly averaged daily solar radiation data from Figure 11-3 are given in Figure 11-4. Best fit curve to data in Figure 11-4 shows that the maximum yearly solar radiation is achieved when a South facing PV collector panel is tilted 48.5 degrees from the horizontal position in Anchorage, AK. Under this tilt angle setting, a PV collector panel in Anchorage, AK will receive 3.08 kWh/(m^2-day) or 1,125 kWh/(m^2-year) of solar radiation which is 45.8% lower than the solar radiation received in San Diego, CA.

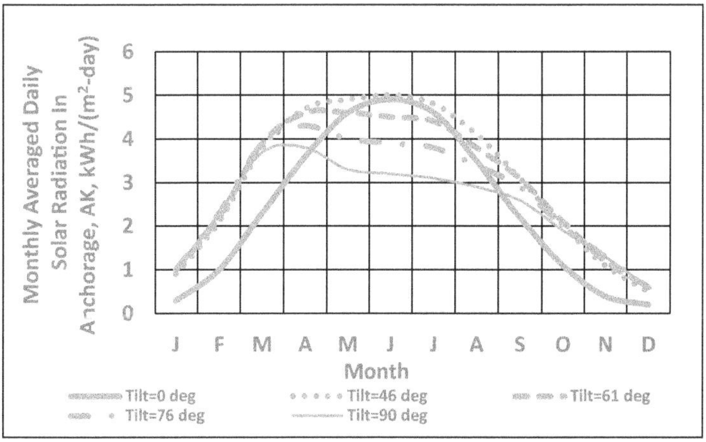

Figure 11-3 Monthly averaged daily solar radiation on a PV collector panel facing South at different tilt angles in Anchorage, AK. at latitude of 61.17 degrees North, longitude of 150.02 degrees West, 35 meters elevation and 1004 millibars mean pressure.

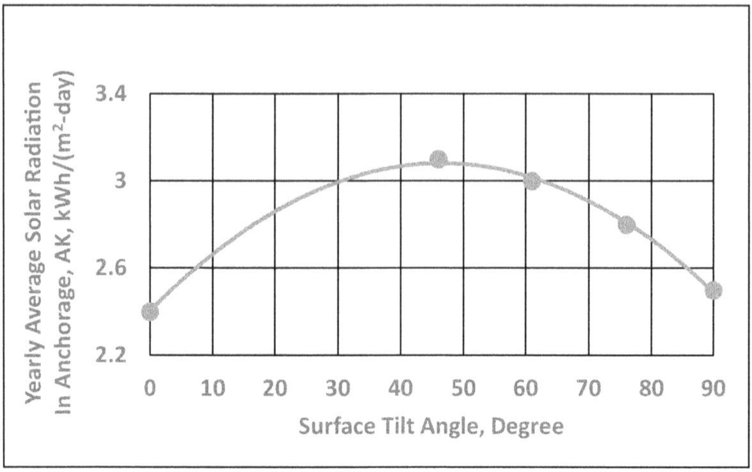

Figure 11-4 Yearly averaged daily solar radiation on a PV collector panel facing South at different tilt angles in Anchorage, AK.

In US, yearly average electricity consumption per household is about 10,399 kWh/yr. Using Equation 11-1, area of solar panels required for this average electricity consumption demand can be determined as shown in Equations 11-2 and 11-3.

For San Diego, CA:

$$A = \frac{10,399}{2,074 \times 0.20 \times 0.70} = 35.8\,m^2 \;\; or \;\; \frac{35.8}{1.5} = 24 \; PV \; collector \; panels \qquad 11\text{-}2$$

For Anchorage, AK:

$$A = \frac{10,399}{1,125 \times 0.20 \times 0.70} = 68.4\,m^2 \;\; or \;\; \frac{68.4}{1.5} = 46 \, PV \; collector \; panels \qquad 11\text{-}3$$

A house in Anchorage, AK has to install almost twice as many PV collector panels as compared to a house in San Diego, CA in order produce the same amount of electrical energy from solar radiation.

Let us next estimate the incident solar radiation on a tilted PV collector panel facing South at two different locations detailed above by using an empirical model given in Ref. [16].

The ratio of solar radiation on a horizontal surface on Earth to the extraterrestrial radiation at the outer edge of Earth's atmosphere at the same time and at the same location is characterized by a stochastic parameter called the clearness index, *K,* as shown in Equation 11-4.

$$\frac{SR_{beam}}{SR_{et}} = K \qquad \text{11-4}$$

where SR_{beam} is the monthly averaged daily direct beam solar radiation on a horizontal surface on Earth in kWh/(m^2-day) and SR_{et} is the extraterrestrial radiation at the outer edge of our atmosphere in kWh/(m^2-day). Both variables are measured at the same time and at the same location. The clearness index characterizes the extinction of solar radiation impinging at the outer edge of our Earth's atmosphere. Extinction of extraterrestrial radiation along a vertical or oblique path in the atmosphere is due to the ozone layer, to other gases such as CO_2, to water vapor (clouds), to aerosols, etc.

The empirical model's results are matched to real solar radiation data given above in Figures 11-1 and Figure 11-3 in order to understand how clearness indices behave throughout the year in San Diego, CA. and in Anchorage, AK.

The estimate for the daily average of extraterrestrial radiation at the outer edge of our atmosphere for a given day and for a given location on Earth is given in Equation 11-5.

$$SR_{et} = \frac{24 \times G \times DAY}{\pi} \times \left[cos(\varphi) cos(\delta) sin(\omega_s) + \left(\frac{\pi \times \omega_s}{180} \right) sin(\varphi) sin(\delta) \right] \qquad \text{11-5}$$

where $G = 1{,}367\ W/m^2$ is the solar constant, DAY is the solar constant day by day variation during the year (see Equation 11-6), φ is the location's latitude angle in degrees, δ is the declination angle in degrees, ω_s sunrise or sunset angle in degrees when the zenith angle of the Sun rays are at - 90 or + 90 degrees respectively, (see Equation 11-7).

$$DAY = \left[1 + 0.033 \times cos\left(\frac{360 \times N}{365} \right) \right] \qquad \text{11-6}$$

where N is the day of the year, i.e. 1 through 365.

$$\omega_s = cos^{-1}[-tan(\varphi) tan(\delta)] \qquad \text{11-7}$$

The declination angle, δ, is the angular position of the solar noon with respect to the plane of the equator as given in Equation 11-8.

$$\delta = 23.45 \times sin\left[\frac{360}{365} (284 + N) \right] \qquad \text{11-8}$$

The declination angle reaches a maximum of +23.45 degrees on June 22 at the summer solstice and a minimum of -23.45 degrees on December 21-22 at the winter solstice in the Northern hemisphere.

If the clearness index in Equation 11-4 can be modeled appropriately, SR_{beam}, the monthly averaged daily direct beam solar radiation on a horizontal surface, can be determined. In order to determine the monthly averaged daily direct beam solar radiation on a tilted surface, SR_{beam}, has to be multiplied by, r_{beam}, which is the ratio of monthly averaged daily direct beam solar radiation on the tilted surface to that on a horizontal surface as given in Equation 11-9.

$$r_{beam} = \frac{(\pi \times \omega_s / 180) sin(\varphi - \beta) sin(\delta) + cos(\varphi - \beta) cos(\delta) sin(\omega_s)}{(\pi \times \omega_S / 180) sin(\varphi) sin(\delta) + cos(\varphi) cos(\delta) sin(\omega_S)} \qquad 11\text{-}9$$

where β is the PV collector panel's tilt angle.

Next, diffuse radiation on a tilted surface has to be determined. If the sky is assumed to be an isotropic source of diffuse radiation, then diffuse radiation falling on a tilted surface can be defined as shown in Equation 11-10.

$$SR_{diffuse} = SR_{diffuse\text{-}horizontal} \times 0.5 \times [1 + cos(\beta)] \qquad 11\text{-}10$$

where $SR_{diffuse\text{-}horizontal}$ is the monthly averaged daily diffuse solar radiation on a horizontal surface on Earth in kWh/(m^2-day).

$SR_{diffuse\text{-}horizontal}$ can be estimated from SR_{beam} which is the monthly averaged daily direct beam solar radiation on a horizontal surface on Earth defined by Equation 11-4, by using Erbs model using cubic polynomials in Equations 11-11a or 11-11b as a function of clearness index K. See Ref. [24].

If $\omega_s \leq 81.4$ *degrees and* $0.3 \leq K \leq 0.8$, then use Equation 11-11a.

$$\frac{SR_{diffuse\text{-}horizontal}}{SR_{beam}} = 1.391 - 3.56K + 4.189K^2 - 2.137K^3 \qquad 11\text{-}11a$$

If $\omega_s > 81.4$ *degrees and* $0.3 \leq K \leq 0.8$, then use Equation 11-11b.

$$\frac{SR_{diffuse\text{-}horizontal}}{SR_{beam}} = 1.311 - 3.022K + 3.427K^2 - 1.821K^3 \qquad 11\text{-}11b$$

Now we can calculate direct beam and diffuse solar radiation components irradiating on a tilted PV collector panel surface using Equations 11-4 through 11-11.

However, there is one more component of solar radiation that has to be considered. It is the ground reflected solar radiation which is also called the ground albedo. The monthly averaged daily ground reflected solar radiation on a tilted surface can be calculated using Equation 11-12.

$$SR_{ground} = (SR_{beam} + SR_{diffuse\text{-}horizontal}) \times \rho \times 0.5 \times [1 - cos(\beta)] \qquad 11\text{-}12$$

where ρ is the ground reflectance and it has a value of about 0.2 for grass, soil and concrete surfaces.

For a tilted surface, total monthly averaged daily total solar radiation on a tilted surface is the sum of direct beam, diffuse and ground reflected components of solar radiation as shown in Equation 11-13.

$$SR_{total} = SR_{beam} \times r_{beam} + SR_{diffuse} + SR_{ground} \qquad 11\text{-}13$$

For a horizontal surface on Earth facing South, Equation 11-13 reduces to Equation 11-14.

$$SR_{total} = SR_{beam} + SR_{diffuse\text{-}horizontal} \qquad 11\text{-}14$$

Monthly averaged daily total solar radiation on a PV collector panel facing South at zero degree tilt angle in San Diego, CA. at latitude of 32.73 degrees North is shown in Figure 11-5. Monthly averaged daily total solar radiation is calculated using Equation 11-14. Monthly averaged clearance index values are adjusted to fit actual data given in Figure 11-1. Monthly averaged clearance index values for zero degree tilt angle calculations for every month are given in Figure 11-9. For San Diego, CA., actual data fit to calculated values at zero degree tilt angle required an yearly averaged clearance index value of 0.411. At zero degree tilt angle, a PV collector panel receives an annual total of 61.1% of the available extraterrestrial solar radiation. Annual direct beam solar radiation is 67.2 % and annual diffuse solar radiation is 32.8 % of the annual total solar radiation that a PV collector panel receives. Figure 11-5 shows total, direct beam, diffuse, extraterrestrial solar radiation and actual data for a horizontal a PV collector panel in San Diego, CA.

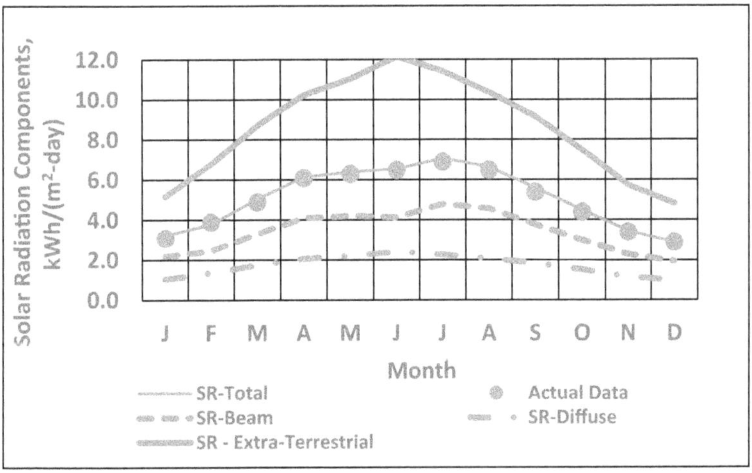

Figure 11-5 Monthly averaged daily total solar radiation and its components on a PV collector panel facing South at zero degree tilt angle in San Diego, CA. at a latitude of 32.73 degrees North.

Monthly averaged daily total solar radiation on a PV collector panel facing South at the optimum tilt angle of 30.07 degrees in San Diego, CA. is shown in Figure 11-6. Monthly averaged daily total solar radiation is calculated using Equation 11-13. Monthly averaged clearance index values are adjusted to fit actual data given in Figure 11-1. A yearly averaged clearance index value of 0.411 is used to fit calculated values to actual data at 30.07 degrees tilt angle. At 30.07 degrees tilt angle, a PV collector panel receives an annual total of 66.7 % of the available extraterrestrial solar radiation. Annual direct beam solar radiation is 70.8 % of the annual total solar radiation that a PV collector panel receives. Annual diffuse solar radiation is 28.0 % and annual ground reflected solar radiation is only 1.2 % of the annual total solar radiation that a PV collector panel receives. Figure 11-5 shows total, direct beam, diffuse, ground reflected, extraterrestrial solar radiation and actual data for a 30.07 degrees tilted PV collector panel in San Diego, CA.

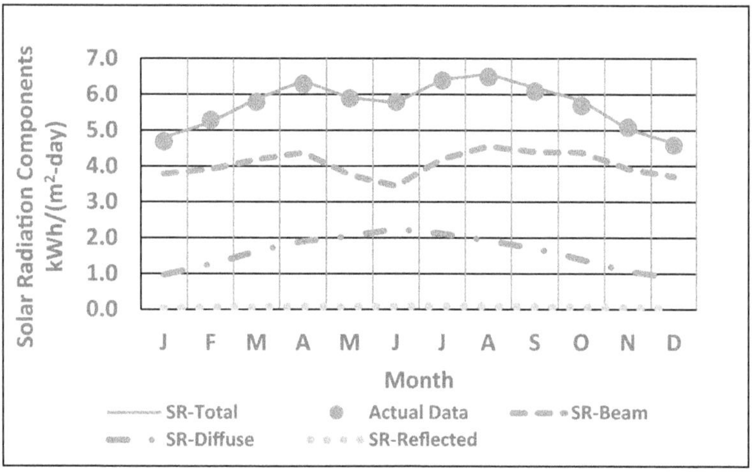

Figure 11-6 Monthly averaged daily total solar radiation and its components on a PV collector panel facing South at the optimum tilt angle of 30.07 degrees in San Diego, CA.

Monthly averaged total daily solar radiation on a PV collector panel facing South at zero degree tilt angle in Anchorage, AK. at latitude of 61.17 degrees North is shown in Figure 11-7. Monthly averaged total daily solar radiation is calculated using Equation 11-14. Monthly averaged clearance index values are adjusted to fit actual data given in Figure 11-3. Monthly averaged clearance index values for zero degree tilt angle calculations for every month are given in Figure 11-9. For Anchorage, AK., actual data fit to calculated values at zero degree tilt angle required an yearly averaged clearance index value of 0.241.

At zero degree tilt angle, a PV collector panel receives 42.1% of the available extraterrestrial solar radiation. Annual direct beam solar radiation is 57.3 % and annual diffuse solar radiation is 42.7 % of the annual total solar radiation that a PV collector panel receives. Figure 11-7 shows total, direct beam, diffuse, extraterrestrial solar radiation and actual data for a horizontal PV collector panel in Anchorage, AK.

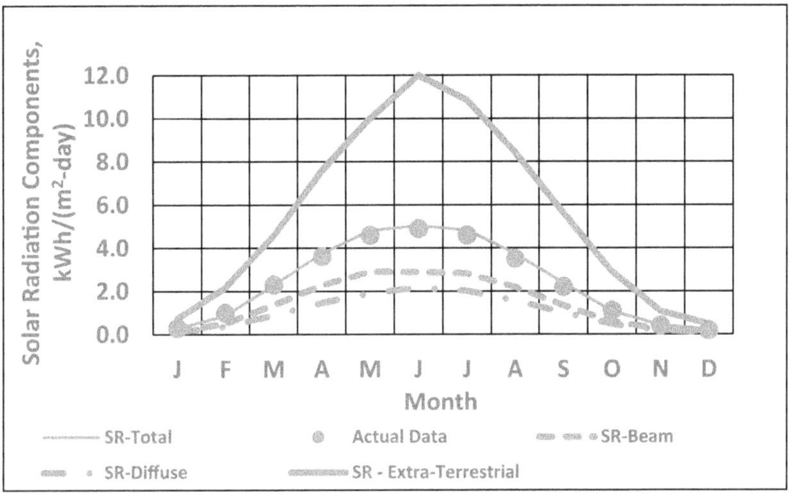

Figure 11-7 Monthly averaged total daily solar radiation and its components on a PV collector panel facing South at zero degree tilt angle in Anchorage, AK. at a latitude of 61.17 degrees North.

Monthly averaged total daily solar radiation on a PV collector panel facing South at the optimum tilt angle of 48.5 degrees in Anchorage, AK. is shown in Figure 11-8. Monthly averaged total daily solar radiation is calculated using Equation 11-13. Monthly averaged clearance index values are adjusted to fit actual data given in Figure 11-3. A yearly averaged clearance index value of 0.241 is used to fit calculated values to actual data. At 48.5 degrees tilt angle. A PV collector panel receives an annual total of 53.7 % of the available extraterrestrial solar radiation. Annual direct beam solar radiation is 69.5 % of the annual total solar radiation that a PV collector panel receives. Annual diffuse solar radiation is 27.9 % and annual ground reflected solar radiation is only 2.6 % of the annual total solar radiation that a PV collector panel receives. Figure 11-8 shows total, direct beam, diffuse, ground reflected, extraterrestrial solar radiation and actual data for a 48.5 degrees tilted PV collector panel in Anchorage, AK.

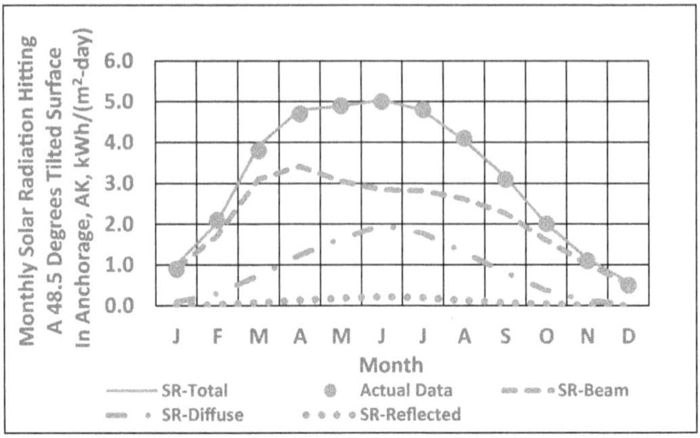

Figure 11-8 Monthly averaged total daily solar radiation on a PV collector panel facing South at the optimum tilt angle of 48.5 degrees in Anchorage, AK.

Monthly averaged clearness indices used to fit present calculations to actual data for a PV collector panel facing South at zero degree tilt angle in San Diego, CA. and in Anchorage, AK. are shown in Figure 11-9. The yearly average value of clearness index for San Diego, CA. is 0.411. The yearly average value of clearness index for Anchorage, AK. is 0.241. These yearly average values of clearness indices can be used to estimate total annual solar radiation on a tilted PV collector panel. It can be very involved undertaking to develop an accurate clearness index model that takes into account all atmospheric effects to predict monthly clearness index values for every location on Earth. Therefore, it is best the fit available appropriate solar radiation data to a solar radiation model such as the one presented in this chapter to optimize PV collector panel tilt angles for a given location.

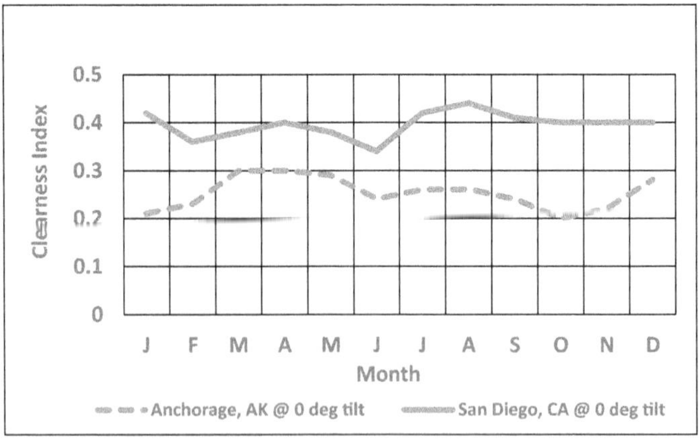

Figure 11-9 Monthly averaged clearness indices used to fit present calculations to actual data for PV collector panels facing South at zero degree tilt angle in San Diego, CA. and in Anchorage, AK.

Yearly averaged extraterrestrial solar radiation, total yearly averaged solar radiation and its components, i.e. direct beam, diffuse and ground reflected, on a PV collector panel facing South with optimized tilted angle are shown for San Diego, CA. and for Anchorage, AK. in Table 11-1.

Table 11-1 Yearly averaged extraterrestrial solar radiation, yearly averaged total solar radiation and its components on a PV collector panel facing South with optimized tilted angles are shown for San Diego, CA. and for Anchorage, AK.164.

	SAN DIEGO, CA.		ANCHORAGE, AK.		
	Optimum Tilt Angle = 30.07 degrees	Percentage Of Extra Terrestrial Radiation	Optimum Tilt Angle = 48.5 degrees	Percentage Of Extra Terrestrial Radiation	Percentage Difference From San Diego, CA.
Extraterrestrial Solar Radiation, kWh/(m^2-year)	3,134	-	2,022	-	-35.5 %
Total Solar Radiation On PV Collector Panel, kWh/(m^2-year)	2,091	66.7 %	1,086	53.7 %	-48.1 %
Direct Beam Solar Radiation On PV Collector Panel, kWh/ (m^2-year)	1,480	47.2 %	754	37.3 %	-49.0 %
Diffuse Solar Radiation On PV Collector Panel, kWh/(m^2-year)	585	18.7 %	302	14.9 %	-48.4 %
Ground Reflected Solar Radiation On PV Collector Panel, kWh/(m^2-year)	26	0.8 %	29	1.4 %	+11.5 %
Yearly Averaged Clearness Index	0.411	-	0.241	-	-41.4 %
NREL Data Used In Present Calculations, kWh/(m^2-year)	2,074	66.2 %	1,125	55.6 %	-45.8 %

12

Converting Solar Radiation to Thermal Energy With a Glazed Flat Plate Collector

There are numerous ways to convert solar radiation to thermal energy in a fluid. In this chapter only glazed flat plate solar radiation collectors will be treated. The function of a solar radiation collector is to absorb solar radiation energy by means of a high absorptivity, i.e. greater than 0.95, surface and convert it into thermal energy. The thermal energy is then transferred from the absorber surface to flowing fluid by means of high thermal conductivity tubes. The collector is protected from convective losses to the environment with a thin film coated glass cover that is highly transparent to incoming short wavelength radiation, i.e. greater than 0.85 transmissivity, but prohibits long wavelength radiation leaving the collector due to its very low emissivity, i.e. less than 0.1. Insulating material is placed along sides and back of the collector in order to minimize convective heat losses to the environment.

Solar radiation to heat energy conversions in a fluid are categorized by being a passive or an active system where a passive system uses natural convection resulting from buoyancy of the heated fluid to store the thermal energy and where an active system uses pumps. A passive or an active system can also be a direct or an indirect converter of thermal energy where a direct system uses the actual fluid such as water or air to carry the available thermal energy and where an indirect system uses a heat exchanger to transfer the thermal energy from a transfer medium such as water-glycol mixture to a medium to be used.

There are several standards that are used to determine the thermal performance of solar collectors. One such standard is ISO (International Organization for Standardization) - 9806:2017 and the other is ASHRAE (American Society of Heating, Refrigeration and Air-Conditioning Engineers) 93-2010.

A detailed heat transfer analysis for converting solar energy into useful thermal energy for a glazed flat plate solar radiation collector can be very involved and complicated. However, a simple heat transfer analysis between the incident solar radiation and the useful thermal energy gained by the transfer fluid can provide us with a good measure of a glazed flat plate collector's performance. The time averaged useful solar energy gain for a glazed flat plate solar radiation collector can be defined as shown in Equation 12-1.

$$Q_{useful} = F \times A \times \left[I_{solar} \times \tau \times \alpha - \left(h_r + h_c \right) \times \left(T_{in} - T_{env} \right) \right] \qquad 12\text{-}1$$

Equation 12-1 is known as the Hottel-Whillier-Bliss equation, see Ref. [29], and it is used in determining a solar radiation collector's performance. Q_{useful} is the time averaged useful thermal energy gained in Watts by the collector that increases transfer medium's temperature. F is the solar radiation collector's heat removal factor. F can vary from 60% to 80% depending on a solar radiation collector's design characteristics. A is the solar radiation collector's area in m^2. I_{solar} is the time averaged total irradiation on the collector in W/m^2. I_{solar} is composed of direct beam radiation, diffuse radiation and ground reflected radiation components. τ is the hemispherical transmissivity averaged over wavelength of total solar radiation through the glazing surface for the collector. a is the hemispherical absorptivity for the collector's absorbing plate averaged over wavelength of total solar radiation that reaches its surface. T_{in} is the heat transfer medium's inlet temperature in K. T_{env} is the environment temperature in K.

h_r is the radiation heat transfer coefficient for radiation emitted in the infrared region from the surface of the solar radiation collector to the sky in $W/(m^2\text{-}K)$ assuming that the whole collector surface is at the transfer medium's inlet temperature, namely T_{in}. h_c is the forced and/or natural convective heat transfer coefficient in $W/(m^2\text{-}K)$ between solar radiation collector's outer surfaces and the environment assuming that the whole collector surface is again at the transfer medium's inlet temperature, namely T_{in}.

The time averaged efficiency, η, of a glazed flat plate solar radiation collector can be defined as the ratio of the time averaged useful energy gained divided by the time averaged incident solar radiation as defined in Equation 12-2.

$$\eta = \frac{Q_{useful}}{A \times I_{solar}} \qquad 12\text{-}2$$

where A is the collector area in m^2. η can be rewritten by combining Equations 12-1 and 12-2 and the result is shown in Equation 12-3.

$$\eta = F \times \tau \times \alpha - F \times \left(h_r + h_c \right) \times \frac{\left(T_{in} - T_{env} \right)}{I_{solar}} \qquad 12\text{-}3$$

In Equation 12-3, time averaged efficiency of a glazed flat plate solar radiation collector is a linear function of $\frac{(T_{in} - T_{env})}{I_{solar}}$ where the maximum efficiency is when $T_{in} = T_{env}$ and it has a value of $F \times \tau \times \alpha$. The collector efficiency has a negative slope of $-F \times (h_r + h_r)$ and η goes to zero when $Q_{useful} = 0$ or when the independent variable $\frac{(T_{in} - T_{env})}{I_{solar}}$ has a value of $\frac{\tau \times \alpha}{(h_r + h_c)}$. When $\eta = 0$, total irradiation coming to the collector equals to the total heat loss from the collector to the environment.

In the present analysis, yearly averaged daily solar radiation data for a fixed flat plate collector in Sacramento, CA at 38.52 degrees North latitude will be used, see Ref. [25].

Yearly averaged daily solar radiation data are given in Figure 12-1 for different flat plate collector tilt angles. Best fit curve to data in Figure 12-1 shows that the maximum yearly solar radiation is achieved when a South facing flat plate collector is tilted 32.6 degrees from the horizontal position in Sacramento, CA. Under 32.6 degrees fixed tilt angle setting, a flat plate collector receives an average of 5.5 kWh/(m^2-day) or 2,008 kWh/(m^2-year) of solar radiation.

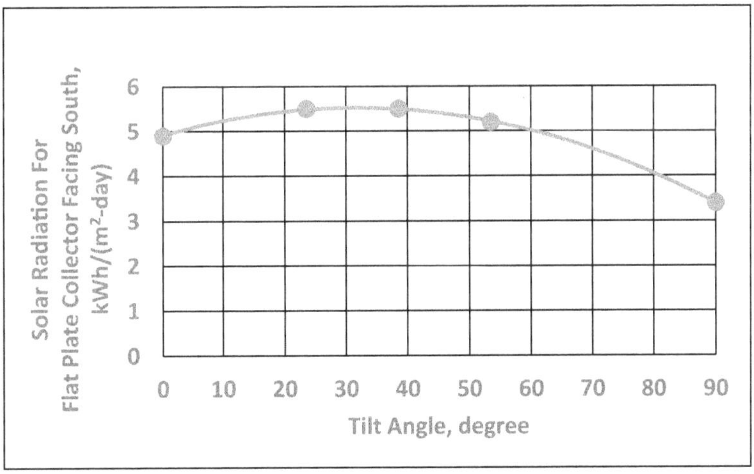

Figure 12-1 Yearly averaged daily solar radiation on a fixed flat plate collector facing South at different fixed tilt angles in Sacramento, CA.

Let us first investigate effects of changing forced and/or natural convective heat transfer coefficient, h_c, to collector efficiency, η, while assuming that the whole collector surface is at the collector's transfer medium's inlet temperature, namely T_{in}. h_c varies from 1 to 10 W/(m^2-K) in the present analysis while holding the following variables as constants.

$I_{solar} = 5.5\ kWh/(m^2 - day)$ from Figure 12-1.

$\tau = 0.9$, $\alpha = 0.94$

$\varepsilon = 0.05$ is the total hemispherical emissivity, namely averaged over all directions and wavelengths in the infrared region, for the thin film coated glass cover of the collector.

$T_{in} = 20\ C\ or\ 293\ K$, $T_{env} = 10\ C\ or\ 283\ K$, $T_{sky} = -20\ C\ or\ 253\ K$, $A = 10\ m^2$, $F = 0.7$.

Q_{useful} in Equation 12-1 can also be equated to temperature rise in the transfer medium's inlet temperature in the collector as shown in Equation 12-4.

$$Q_{useful} = \dot{V} \times \rho \times c_p \times (T_{out} - T_{in}) \qquad 12\text{-}4$$

where $\dot{V}$ is the heat transfer medium's volumetric flow rate in liters/day. ρ is the density of the heat transfer medium in kg/m^3. c_p is the specific heat at constant pressure of the heat transfer medium in J/(kg-K). T_{out} is the heat transfer medium's outlet temperature from the collector in K. In the present analysis, following values are used for parameters in Equation 12-4.

$$\dot{V} = 864\ liters/day,\ \rho = 1{,}000\ kg/m^3,\ c_p = 4{,}184\ J(kg - K)$$

Using above parameters and Equation 12-3, effects of varying convective heat transfer coefficient, h_c, on collector's efficiency are shown in Figure 12-2. With increasing wind velocity over the collector's surfaces, the convective heat transfer coefficient increases and so does the negative slope of the collector efficiency versus $\frac{(T_{in} - T_{env})}{I_{solor}}$ line.

With increasing h_c, the collector efficiency decreases faster as expected.

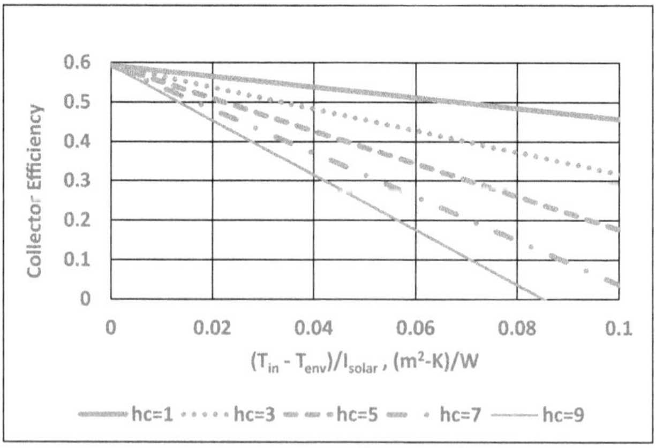

Figure 12-2 Collector efficiency versus $\frac{(T_{in} - T_{env})}{I_{solor}}$ for increasing h_c values.

Slope of the linear relationship in Equation 12-3, namely $F \times (h_r + h_c)$, versus the convective heat transfer coefficient is shown in Figure 12-3. Sensitivity of the slope's value to the convective heat transfer coefficient's value is - 0.7 and it is a constant. The slope's value decreases by 70% of the positive change in the value of the convective heat transfer coefficient which is the collector's heat removal factor in the present analysis. Radiation emitted in the infrared region from the surface of the collector to the sky does not affect the slope changes in the present analysis.

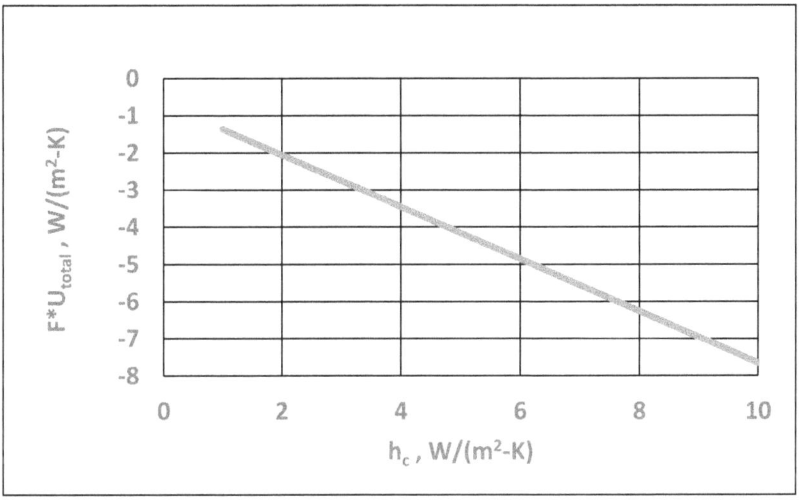

Figure 12-3 Slope of the linear relationship in Equation 12-3, namely $F \times (h_r + h_c)$, versus the convective heat transfer coefficient.

Figure 12-4 presents the percent decrease in collector's efficiency when the initial value of h_c is doubled at different $\frac{(T_{in} - T_{env})}{I_{solor}}$ values. When the solar collector is operating at $\frac{(T_{in} - T_{env})}{I_{solor}} = 0.02$ and $h_c = 5$ W/(m^2-K), and if the value of h_c doubles, then the efficiency of the collector drops by 13.7%. However, if the solar collector is operating at $\frac{(T_{in} - T_{env})}{I_{solor}} = 0.08$ and $h_c = 5$ W/(m^2-K), and if the value of h_c doubles, then the efficiency of the collector drops by 107.6%.

Figure 12-5 shows the heat transfer medium's outlet temperature, T_{out}, versus convective heat transfer coefficient by using Equation 12-4. The heat transfer medium's outlet temperature decreases linearly with increasing convective heat transfer coefficient when all other parameters are constants as given above. The heat transfer medium's outlet temperature decreases 1.67 times the increase in convective heat transfer coefficient.

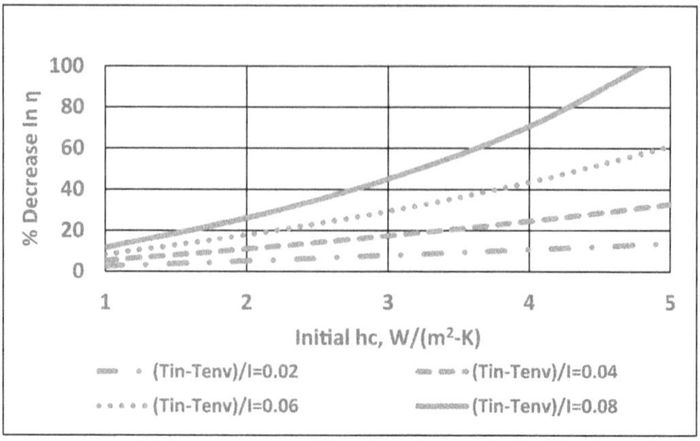

Figure 12-4 Percent decrease in collector's efficiency when the initial value of h_c is doubled at different $\frac{(T_{in}-T_{env})}{I_{solor}}$ values.

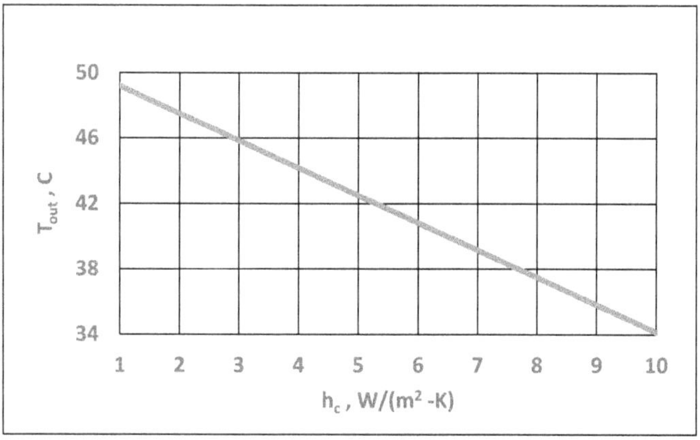

Figure 12-5 Heat transfer medium's outlet temperature versus convective heat transfer coefficient.

Using above constant parameters, $h_c = 3.0\ W/(m^2 - K)$ and Equation 12-3, effects of varying total irradiation, I_{solar}, on collector's efficiency are shown in Figure 12-6. In the present analysis, the maximum collector efficiency is 0.59, namely $F \times \tau \times \alpha = 0.59$ and the slope of the linear line is - 2.75, namely $-F \times (h_r + h_c) = -2.75$ are used.

The monthly averaged daily total irradiation, I_{solar}, for Sacramento, California varies from a minimum of 2.7 in December to a maximum of 7.2 in July and in August in kWh/(m²-day), see Ref. [25]. In July and in August, when $I_{solar} = 7.2\ kWh/(m^2 - K)$, the collector efficiency climbs up to 0.501. In December, when $I_{solar} = 2.7\ kWh/(m^2 - K)$, the collector efficiency drops down to 0.348.

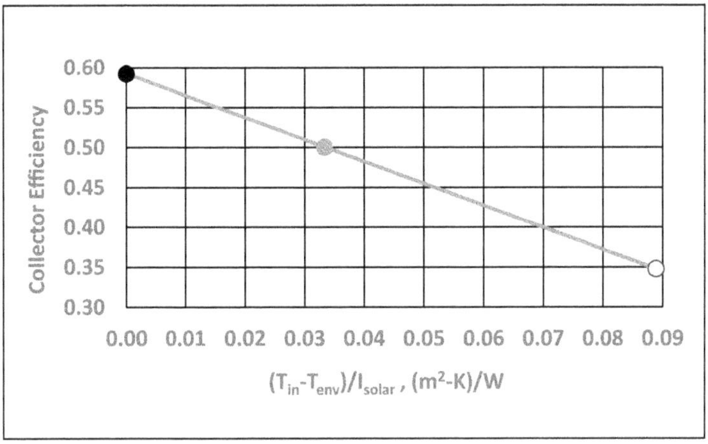

Figure 12-6 Collector efficiency versus varying monthly averaged daily total irradiation, I_{solar}.

In Figure 12-6, black dot corresponds to maximum collector efficiency of 0.59, gray dot corresponds to July and August irradiation, when $I_{solar} = 7.2\ kWh/(m^2 - K)$ and white dot corresponds to December irradiation, when $I_{solar} = 2.7\ kWh/(m^2 - K)$.

Sensitivity of collector efficiency to changes in monthly averaged daily total irradiation, I_{solar}. is shown in Figure 12-7. Sensitivity is high at low irradiation values and it decreases as irradiation values increase, namely the sensitivity decreases as I_{solor}^{-2}.

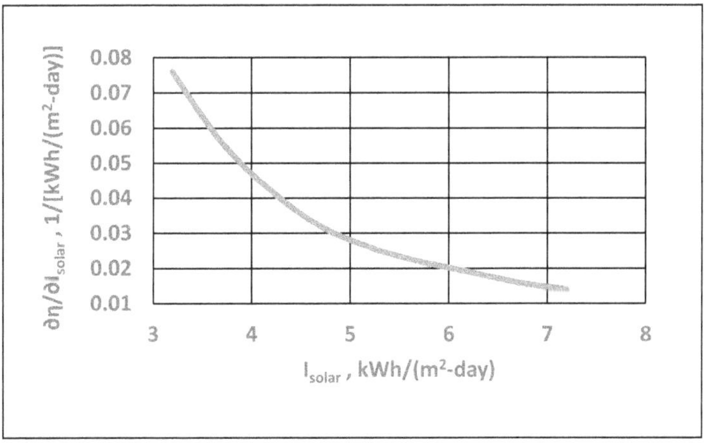

Figure 12-7 Sensitivity of collector efficiency to changes in monthly averaged daily total irradiation, I_{solar}.

Figure 12-8 shows the collector efficiency versus $\frac{(T_{in} - T_{env})}{I_{solor}}$ for different values of heat transfer medium's inlet temperature, T_{in}. As T_{in} increases, the temperature potential between the collector and the environment increases which also increases the heat loss from the collector to the environment. Equation 12-1 states that useful heat, Q_{useful}, coming into the collector decreases and therefore the collector efficiency decreases as shown in Figure 12-8. Black dot in Figure 12-8 represents the maximum collector efficiency of 0.59 at $T_{in} - T_{env} = 0$. White dot is for $T_{in} - T_{env} = 1\ C$ with an efficiency value of 0.58. Gray dot is for $T_{in} - T_{env} = 3\ C$ with an efficiency value of 0.56. Black triangle is for $T_{in} - T_{env} = 5\ C$ with an efficiency value of 0.54. White triangle is for $T_{in} - T_{env} = 10\ C$ has an efficiency value of 0.49 which is a 16.7% drop in the collector efficiency from its maximum value.

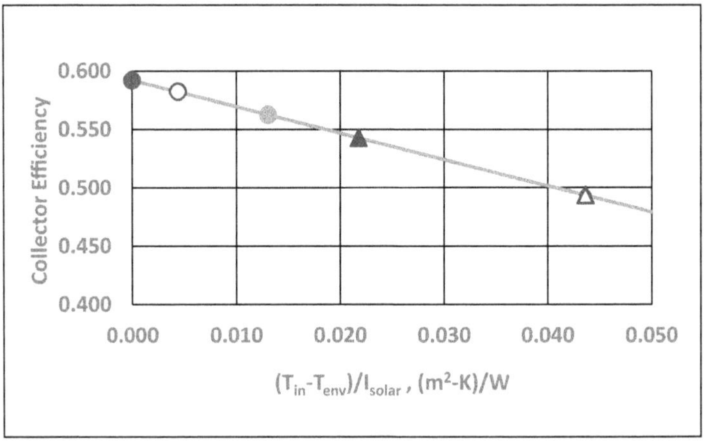

Figure 12-8 Collector efficiency versus $\frac{(T_{in} - T_{env})}{I_{solor}}$ for different values of $T_{in.}$

Effects of changing emissivity, ε, for the thin film coated surface of the glass cover for the collector can also affect the thermal efficiency of the collector. Figure 12-9 shows the thermal efficiency of the collector versus $\frac{(T_{in} - T_{env})}{I_{solor}}$ for different values of emissivity. Radiation emitted in the infrared region from the glass surface of the collector to the sky increases linearly with increasing emissivity. This increase in radiation heat loss also increases the slope of the thermal efficiency curve as shown in Figure 12-9. If emissivity increases from 0.01 to 0.10, the absolute value of the slope increases by 52.4% from 2.23 W/(m²-K) to 3.40 W/(m²-K).

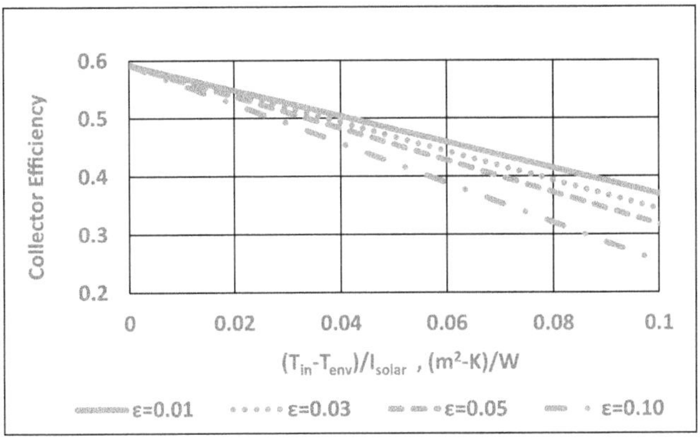

Figure 12-9 Thermal efficiency of the collector versus $\frac{(T_{in} - T_{env})}{I_{solor}}$ for different values of total hemispherical emissivity, ε, for the thin film coated surface of the glass cover for the collector.

Next let us investigate effect of changing glazing surface transmissivity and absorbing plate's absorptivity on collector's thermal efficiency. Figure 12-10 shows collector's thermal efficiency versus $\frac{(T_{in} - T_{env})}{I_{solor}}$ for different values of $\tau \times \alpha$, namely for the product of the transmissivity of the glazing surface and the absorptivity of the absorbing plate for the collector. $\tau \times \alpha$ affects the maximum value of the thermal efficiency. Changing $\tau \times \alpha$ value moves the thermal efficiency curve up or down parallel to the original line depending upon the change ratio. For example, there is a 50% increase in the value of $\tau \times \alpha$ from 0.6 to 0.9 in Figure 12-10.

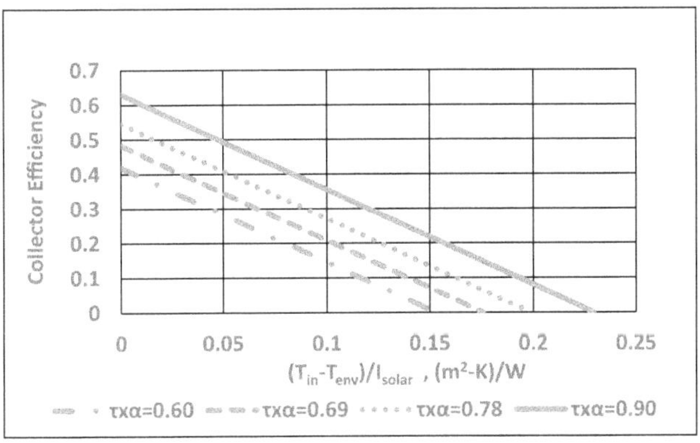

Figure 12-10 Collector efficiency versus $\frac{(T_{in} - T_{env})}{I_{solor}}$ for different values of $\tau \times \alpha$, namely for the product of the transmissivity of the glazing surface and the absorptivity of the absorbing plate for the collector.

Increasing $\tau \times \alpha$ value increases the incoming irradiation on the collector and therefore increases the useful heat going into the collector. Figure 12-11 shows the heat transfer medium's outlet temperature versus $\tau \times \alpha$ for $T_{in} = 20$ *C* and for $V = 864$ *liters/day.*

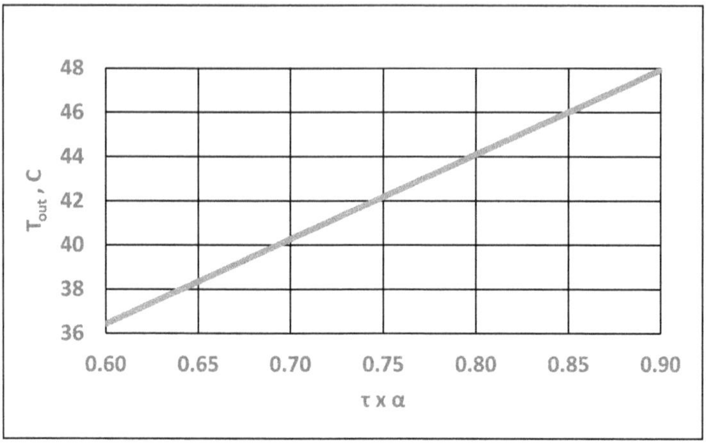

Figure 12-11 Heat transfer medium's outlet temperature versus $\tau \times \alpha$ for $T_{in} = 20$ *C.*

13

Human Thermal Comfort

Humans are very sensitive to changes in environmental temperature and humidity because our internal body temperature has to be kept at a narrow temperature range of 37 ± 0.5 C. Hypothalamus is the central control organ in the brain that controls our internal body temperature. When our internal body temperature rises above 37.5 C, our blood flow is increased to our skin in order to increase the heat transfer from our skin to the environment. At elevated internal body temperature sweating also starts and cools our skin which increases the heat transfer from the body core to our skin. When our internal body temperature falls below 36.5 C, our blood flow is decreased to our skin in order to decrease the heat transfer from our skin to the environment. Also when our internal body temperature falls below 36.5 C, muscle shivering can start in order to increase the internal metabolic heat generation. Thermal energy balance between a human body and the environment is shown in Equation 13-1.

$$M - W = \left(Q_c + Q_r + E_e\right)_{skin} + \left(Q_{cl} + E_{el}\right)_{lungs} + \left(S_{ss} + S_{cs}\right) \qquad 13\text{-}1$$

where M is the metabolic heat generation in Watts. For an average adult male at rest with a skin surface area of 1.8 m^2, M = 100 *W*. If shivering occurs, heat generated by shivering should be added to metabolic heat generation. W is the external work done by our muscles. Q_c is the convection heat transfer from the skin surface to the environment. Q_r is the radiation heat transfer from the skin surface to surroundings. E_e is the latent heat loss due to evaporation on the skin which depends on the relative humidity of the environment and on skin's wetness. Q_{cl} is the convection heat loss from respiration. E_{el} is the latent heat loss from respiration. S_{ss} is the energy stored in the skin and S_{cs} is the energy stored in body's core.

When we have clothing on, heat transfer calculations get more complicated. Convection and radiation heat losses which are also called sensible heat losses, to the environment and to surroundings are from the surface area of the clothing, A_{clo} in m^2, which is at temperature T_{clo} in K, to the environment at T_{env} in K as given in Equation 13-2.

$$Q_c + Q_r = A_{clo} \times (h_c + h_r) \times (T_{clo} - T_{env}) \qquad 13\text{-}2$$

where h_c is the convection heat transfer coefficient in W/(m^2-K) and h_r is the radiation heat transfer coefficient in W/(m^2-K) and it is given in Equation 13-3.

$$h_r = \frac{\varepsilon \times \sigma \times \left(T_{clo}^4 - T_{surr}^4\right)}{\left(T_{clo} - T_{env}\right)} \qquad 13\text{-}3$$

Equation 13-3 assumes a view factor of unity between the clothing and the surroundings. ε is the hemispherical emissivity of the clothing surface averaged over all directions and wavelengths. σ is the Stefan-Boltzmann constant, namely 5.67×10^{-8} W/(m^2-K^4). T_{surr} is the average surrounding temperature in K.

Under steady-state conditions, sensible heat loss from the skin to the clothing surface is given in Equation 13-4. Heat transferred from the skin's surface, A_{sk} in m^2, through the clothing by conduction, convection and radiation or through a total clothing resistance of R_{clo} in (m^2-K)/W as shown in Equation 13-4.

$$Q_c + Q_r = \frac{A_{sk} \times (T_{sk} - T_{clo})}{R_{clo}} \qquad 13\text{-}4$$

Heat loss through clothing is determined experimentally and it is published in I_{clo} values for different clothings. See Ref. [1]. Relationship between R_{clo} and I_{clo} is given in Equation 13-5.

$$R_{clo} = 0.155 \times I_{clo} \qquad 13\text{-}5$$

A_{sk} is the body skin surface area proposed by Du Bois, see Ref. [1], and it is given in Equation 13-6.

$$A_{sk} = 0.202 \times W^{0.425} \times H^{0.725} \qquad 13\text{-}6$$

where W is the weight of the person in kg and H is the height of the person in m.

As we wear clothing our surface area enlarges. f_{clo} is the clothing area factor which is the ratio of A_{clo} to A_{sk} and it is estimated by Equation 13-7. See Ref. [1].

$$f_{clo} = 1.0 + 0.3 \times I_{clo} \qquad 13\text{-}7$$

Sensible heat loss from skin surface to the environment is obtained by combining Equations 13-2 and 13-4 and getting rid of T_{clo} as shown in Equation 13-8.

$$Q_c + Q_r = \frac{A_{sk} \times (T_{sk} - T_{env})}{R_{clo} + \dfrac{1}{f_{clo} \times (h_c + h_r)}} \qquad 13\text{-}8$$

For the following example calculations, h_c in W/(m^2-K) is given in Equation 13-9, for a standing person in moving air. See Ref. [1].

$$
\begin{array}{ll}
0 < V \le 0.15 & h_c = 4.0 \\
0.15 < V \le 1.5 & h_c = 14.8 \times V^{0.69}
\end{array}
\qquad 13\text{-}9
$$

where V is the velocity of air in m/s. Also the non-linear relationship between radiation heat transfer coefficient and temperature given in Equation 13-3 is approximated by the following Equation 13-10. See Ref. [1].

$$h_r = 4.7 \times \varepsilon \qquad 13\text{-}10$$

In following calculations clothing surface is assumed to be a blackbody radiation emitter and therefore $\varepsilon = 1.0$. At low air velocities, i.e. $V = 0.15\frac{m}{s}$ *or* 0.54 *km/hr;* magnitudes of h_c and h_r have equal magnitudes. However, as the air velocity increases, h_c becomes dominant. When $V = 0.15\frac{m}{s}$ *or* 5.4 *km/hr;* $h_c/h_r = 4.2$.

Sensible heat loss from the skin surface to the environment is analyzed for a special case by using Equation 13-8. It is assumed that the metabolic heat generation of a person standing relaxed and weighing 80 kg with a height of 1.8 meters is low, i.e. less than 100 W. The person is standing in low humidity moving air environment so that latent heat loss from the skin is low and heat loss from respiration is low, i.e. less than 10%, as compared to convection and radiation heat losses from the skin. For thermal comfort, the skin temperature is set by the person's thermal controls at 33 C.

In the first case, the person is wearing shorts and a short sleeve shirt which has an I_{clo} value of 0.36 in a 23 C environment. In low air velocity where $h_c = 4.0\ W/(m^2 - K)$, this person's sensible heat loss to the environment with clothing is 119.8 W. Same person without clothing, i.e. $R_{clo} = 0$, and for a similar thermal comfort, needs an environment at a higher temperature of $T_{env} = 25.8\ C$, namely a lower potential difference of 7.2 C instead of 10 C between skin and environment temperatures in order to compensate the lower total resistance of sensible heat flow between the skin and the environment. With this light clothing, if the clothed environment temperature is increased from 23 C to 30 C without increasing latent heat and respiration losses, for a similar thermal comfort, the environment temperature for the unclothed person has to increase linearly by 0.72 C for every degree of increase in the light clothed person's environment temperature. For similar thermal comfort, i.e. $T_{sk} = 33\ C$, temperature of the environment without clothing versus with light clothing is shown in Figure 13-1.

As the environment temperature increases, sensible heat losses from the skin to the environment decreases as shown in Figure 13-2. Sensible heat loss decreases linearly by - 12 W/C since the potential temperature difference also decreases linearly.

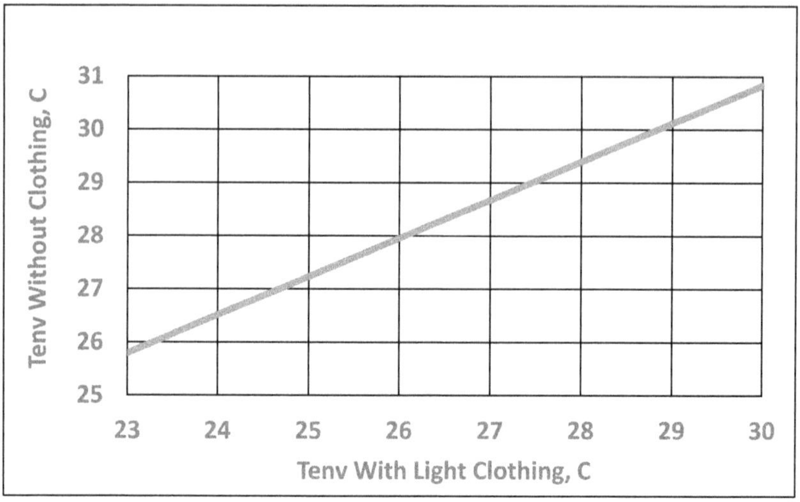

Figure 13-1 Temperature of environment without clothing versus with light clothing for thermal comfort.

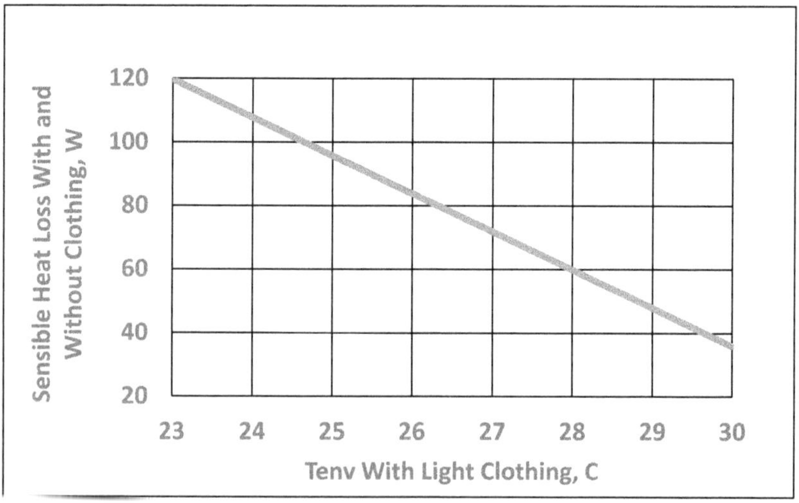

Figure 13-2 Sensible heat loss with light clothing and without clothing versus environment temperature with light clothing.

Above analysis for thermal comfort can also be performed by varying air velocity around the person from 0.54 km/hr (0.15 m/s) to 5.4 km/hr (1.5 m/s) using Equation 13-9. The light clothed person is wearing shorts and a short sleeve shirt which has an I_{clo} value of 0.36 in a 23 C environment. In low air velocity of 0.54 km/hr, i.e. $h_c = 4.0\ W/(m^2 - K)$, this person's sensible heat loss to the environment with clothing is 119.8 W.

For thermal comfort without clothing, the environment temperature increases non-linearly as air velocity increases as shown in Figure 13-3. Sensitivity of the environment temperature without clothing to increasing air velocity is shown in Figure 13-4. Sensitivity is high at low air velocities and decreases as the air velocity increases.

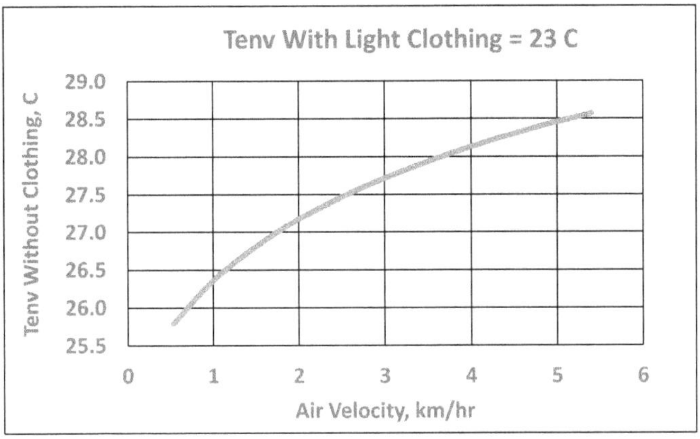

Figure 13-3 Temperature of environment without clothing versus air velocity for thermal comfort.

As the air velocity increases so does the convective heat transfer coefficient. In order to have the same thermal comfort for the unclothed person, temperature of the environment increases to compensate for both $R_{clo} = 0$ and for increasing convection heat transfer coefficient. These two changing heat transfer parameters affect the unclothed person's environment temperature sensitivity to changing air velocity nonlinearly. Sensitivity of a change in environment temperature for the unclothed person to a change in air velocity versus air velocity is shown in Figure 13-4.

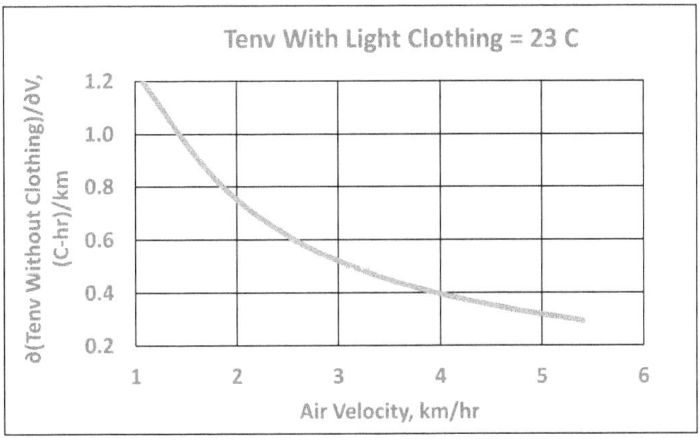

Figure 13-4 Sensitivity of environment temperature without clothing to increasing air velocity.

Sensible heat loss also increases with increasing air velocity for the light clothed person at T_{env} = 23 *C* in order to keep the skin temperature at 33 C. Sensible heat loss versus air velocity is shown in Figure 13-5 for the light clothed person. Sensitivity of sensible heat loss to air velocity versus air velocity is shown in Figure 13-6. This sensitivity behavior in Figure 13-6 is similar to the environmental temperature without clothing sensitivity behavior shown in Figure 13-4.

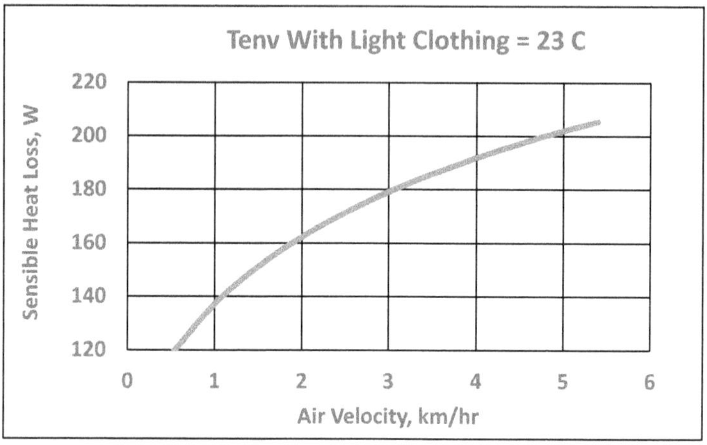

Figure 13-5 Sensible heat loss versus air velocity for the light clothed person.

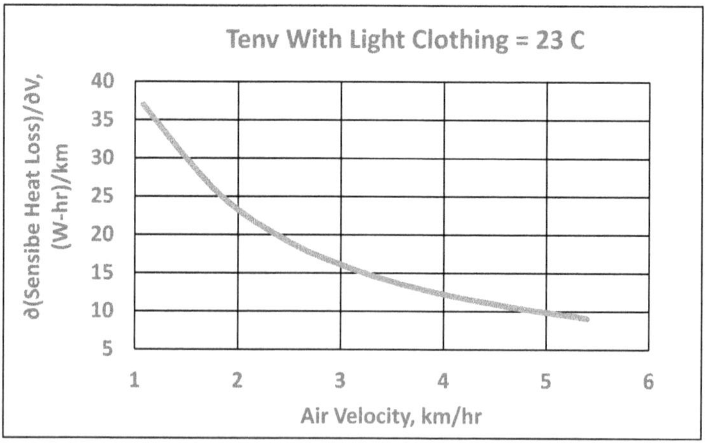

Figure 13-6 Sensitivity of sensible heat loss to increasing velocity for the light clothed person.

In the second case, the same person for a thermal comfort at T_{sk} = 33 *C*, is wearing a long sleeved shirt, a long sleeved sweater, a suit jacket and trousers, a t-shirt and a long underwear bottom which has an I_{clo} value of 1.30 in a 10 C environment. In low air velocity, where $h_c = 4.0\ W/(m^2 - K)$, this person's convection and radiation heat loss to the environment with heavy clothing is 154.7 W using Equation 13-8.

Same person without clothing, i.e. $R_{clo} = 0$, and for a similar thermal comfort, needs an environment at a higher temperature of $T_{env} = 23.7\ C$, namely a lower potential difference of 9.3 C instead of 13 C between skin and environment temperatures in order to compensate lower resistance of sensible heat flow between the skin and the environment. With this heavy clothing, if the clothed environment temperature is increased from 10 C to 17 C without increasing latent heat and respiration losses, for a similar thermal comfort, the environment temperature for the unclothed person has to increase linearly by 0.40 C for every degree of increase in the heavy clothed person's environment temperature. This linear sensitivity of 0.40 C for heavy clothed person is 44% of the linear sensitivity for light clothed person because of 4X R_{clo} for the heavy clothed person. For similar thermal comfort, i.e. $T_{sk} = 33\ C$, temperature of the environment without clothing versus with clothing is shown in Figure 13-7.

As the environment temperature increases, sensible heat loss from the skin to the environment decreases as shown in Figure 13-8. Sensible heat loss decreases linearly by - 6.7 W/C for the heavy clothed person as compared to - 12.0 W/C for the light clothed person since total sensible heat transfer resistance almost doubles for the heavy clothed person at 0.284 (m^2-K)/W as compared to the light clothed person at 0.160 (m^2-K)/W.

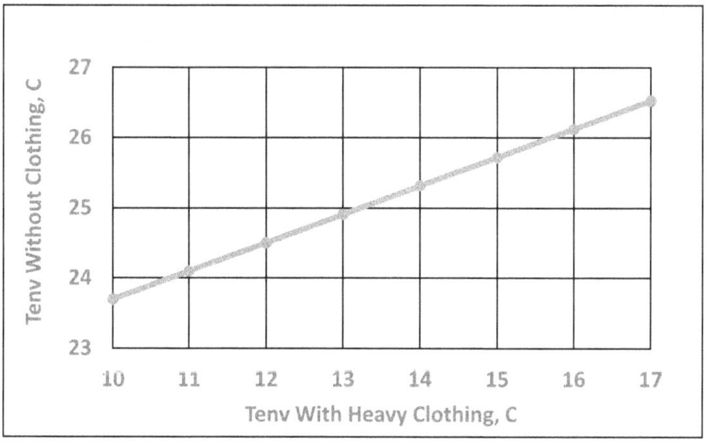

Figure 13-7 Temperature of environment without clothing versus with heavy clothing for thermal comfort.

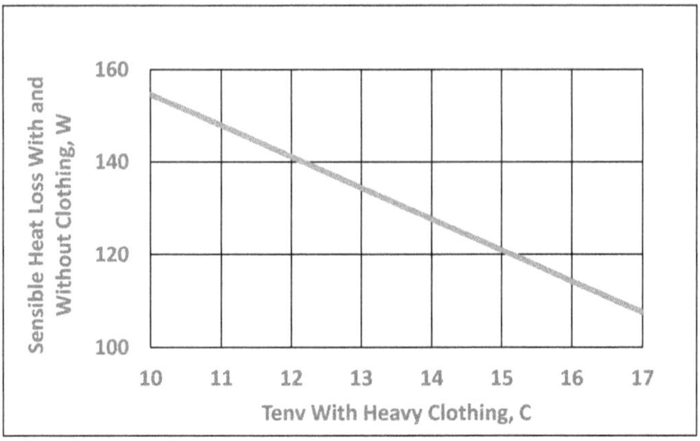

Figure 13-8 Sensible heat loss with heavy clothing and without clothing versus environment temperature with heavy clothing.

Above analysis for thermal comfort for the heavy clothed person can also be performed by varying air velocity around the heavy clothed person from 0.54 km/hr (0.15 m/s) to 5.4 km/hr (1.5 m/s) using Equation 13-9. The heavy clothed person has an I_{clo} value of 1.30 in a 10 C environment. In low air velocity of 0.54 km/hr, $h_c = 4.0\ W/(m^2 - K)$, this heavy clothed person's sensible heat loss to the environment is 154.7 W.

For thermal comfort, the environment temperature increases non-linearly as air velocity increases as shown in Figure 13-9. Sensitivity of the environment temperature without clothing to increasing air velocity is shown in Figure 13-10. Sensitivity is high at low air velocities and decreases as the air velocity increases. Sensitivity values are almost doubles for the heavy clothed person as compared to the light clothed person.

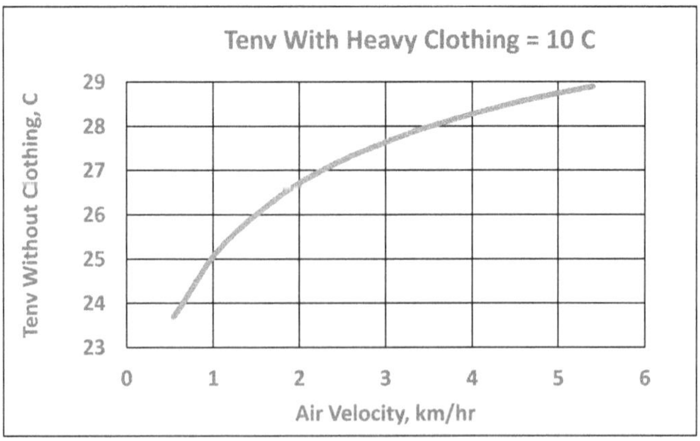

Figure 13-9 Temperature of environment without clothing versus air velocity for thermal comfort.

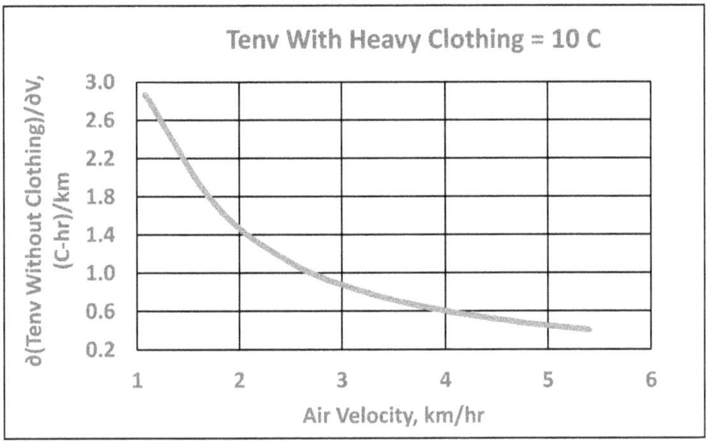

Figure 13-10 Sensitivity of environment temperature without clothing to increasing air velocity.

Sensible heat loss also increases with increasing air velocity for the heavy clothed person at T_{env} = 10 C in order to keep the skin temperature at 33 C. Sensible heat loss versus air velocity is shown in Figure 13-11 for the heavy clothed person. Sensitivity of sensible heat loss to air velocity versus air velocity is shown in Figure 13-12. These sensitivity values for the heavy clothed person are half the sensitivity values for the light clothed person shown in Figure 13-6 since R_{clo} dominates the sensible heat flow resistances from the skin to the environment for the heavy clothed person.

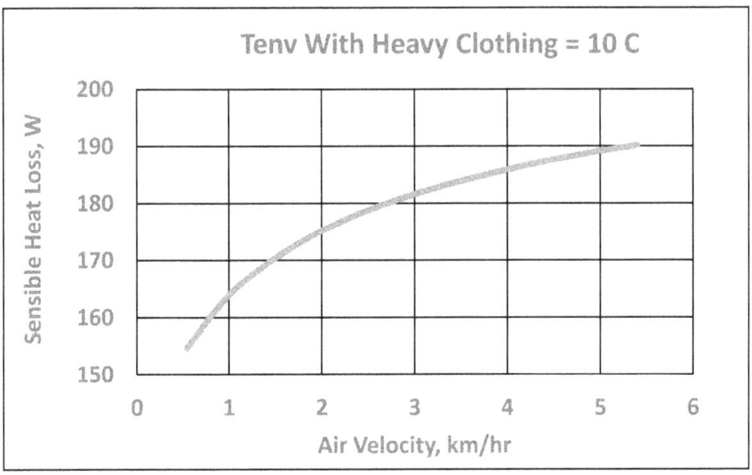

Figure 13-11 Sensible heat loss versus air velocity for the heavy clothed person.

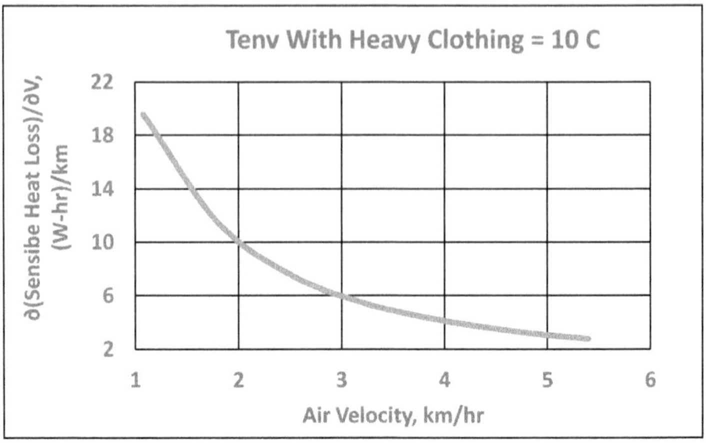

Figure 13-12 Sensitivity of sensible heat loss to increasing velocity for the heavy clothed person.

14

Energy Balance for a Swimming Pool

Swimming pools use solar collectors or natural gas heaters to maintain comfortable temperatures that are desired, i.e. between 20 C to 30 C. UV stabilized transparent plastic or bubble type covers are used during night to minimize losses due to evaporation heat transfer and convection heat transfer from pool water's surface. Since water in a pool has a very large thermal mass, water's temperature changes slowly. Therefore in the present analysis, steady-state heat transfer conditions will be analyzed. Also monthly averaged solar radiation absorbed by water, relative humidity of air, latent heat of evaporation and diffusion coefficient of water in air will be used. Conduction heat transfer from pool walls to earth will be neglected, since this mode of heat transfer is less than 2% as compared to major heat transfer mechanisms, such as incident solar radiation, evaporation from surface of pool water, convection losses from surface of pool water and infrared radiation losses from surface of pool water. With above assumptions, net heat loss from surface of pool water in a month can be estimated using Equation 14-1.

$$Q_{loss} = Q_{rad} + Q_{conv} + Q_{evap} - Q_{solar} \qquad 14\text{-}1$$

where Q_{loss} is net heat loss from water in the pool in a month in GJ/month. Q_{conv} is convection heat loss from water's surface to air in a month in GJ/month. Q_{evap} is heat loss due to evaporation from water's surface in a month in GJ/month. Q_{solar} is the monthly incident total solar radiation, including direct, diffuse and ground reflected radiation, in GJ/month.

Infrared radiation loss from surface of water to sky in GJ/month is given in Equation 14-2.

$$Q_{rad} = 0.002592 \times A \times \sigma \times \varepsilon \times \left(T_w^4 - T_{sky}^4\right) \qquad 14\text{-}2$$

where $A = L \times W$ is the top surface area of a rectangular pool in m^2, (L is the length of the pool in m and W is the width of the pool in m). σ is the Stefan-Boltzmann constant, namely 5.67×10^{-8} W/(m^2-K^4). ε is the hemispherical emissivity of water's surface averaged over all directions and wavelengths. In following calculations, it is assumed that ε = 1.0 for water's surface with or without a cover. T_{sky} is the average sky temperature in K and it is assumed to have a monthly average value of 273 K for the following calculations. The constant value of 0.002592 is the conversion factor from Watts to GJ/month.

Convection heat transfer from top surface area of water is obtained using Equation 14-3 in GJ/month.

$$Q_{conv} = 0.002592 \times A \times h_c \times (T_w - T_a) \qquad \text{14-3}$$

where h_c is the average convection heat transfer coefficient for the entire length of the pool with a laminar to turbulent transition at Reynolds number of 5×10^5. Length averaged convection heat transfer coefficient is given in Equation 14-4.

$$Nu = \left(\frac{h_c \times L}{k}\right) = \left(0.037 \times Re_L^{0.8} - 871\right) \times Pr^{0.333} \qquad \text{14-4}$$

where Nu is the Nusselt number that represents the dimensionless temperature gradient above water's surface. k is the thermal conductivity of air in W/(m-K). $Re_L = \frac{U \times L}{\nu}$ is the dimensionless Reynolds number that is the ratio of inertia to viscous forces of flowing air in the boundary layer. U is the free stream velocity of air in m/s. $\nu = \frac{\mu}{\rho}$ is the kinematic viscosity of air in m^2/s where μ. is the viscosity of air in (N-s)/m^2 and ρ is the density of air in kg/m^3. $Pr = \frac{\nu}{\alpha}$ is the Prandtl number of air which is the ratio of kinematic viscosity and thermal diffusivity of air. $\alpha = \frac{k}{\rho \times c_p}$ where c_p is the specific heat of air at constant pressure in (W-hr)/(kg-K). In this study, all above properties of air are taken at film temperature which is the average of pool's water temperature and the air temperature. Equation 14-4 is valid for $5 \times 10^5 < Re_L < 1 \times 10^8$ and for $0.6 < Pr < 60$. See Ref. [13].

Heat lost from pool water surface to air due to evaporation is given in Equation 14-5 in GJ/month.

$$Q_{evap} = 0.002592 \times M \times H_{fg} \qquad \text{14-5}$$

where M is the evaporation rate in kg/(m^2-hr) and H_{fg} is the latent heat of evaporation of water in air in (W-hr)/kg. The evaporation rate at the boundary of water and air can be defined in terms of a convection mass transfer coefficient similar to the convection heat transfer coefficient and also by assuming the water vapor as a perfect gas is shown in Equation 14-6.

$$M = h_m \times \frac{(p_w - RH \times p_a)}{R \times T_w} \qquad 14\text{-}6$$

where h_m is the mass transfer coefficient at the boundary layer between pool's water surface and air in m/hr. p_w is the saturation pressure of water vapor at pool's water temperature in N/m^2. p_a is the saturation pressure of water vapor at air temperature in N/m^2. *RH* is the average relative humidity for the month which has a value of 0.50 in the present calculations. *R* is the gas constant for water vapor with the value of 461.9 (N-m)/(kg-K). Saturation pressures for water vapor in Pascal are obtained for a particular surrounding temperature *T* in C from Equation 14-7 (August-Roche-Magnus formula) which comes from a best fit empirical equation to experimental data.

$$p_{sat} = 610.94 \times exp\left(\frac{17.62 \times T}{T + 243.04}\right) \qquad 14\text{-}7$$

The mass transfer coefficient can be obtained by assuming that the water in air concentration gradient at the boundary layer of water-air interface is the same as the temperature gradient, namely $Nu = Sh = \frac{h_m \times L}{D}$. *D* is the diffusion coefficient of water in air in m^2/hr. *Sh* is the Sherwood number and it can be written similar to Equation 14-4 as shown in Equation 14-8.

$$Sh = \left(\frac{h_m \times L}{D}\right) = \left(0.037 \times Re_L^{0.8} - 871\right) \times Sc^{0.333} \qquad 14.8$$

where $Sc = \frac{\nu}{D}$ is the Schmidt number, i.e. the ratio of kinematic viscosity to diffusion coefficient of water in air. Equation 14-8 is valid for $5 \times 10^5 < Re_L < 1 \times 10^8$ and for $0.6 < Sc < 3{,}000$. See Ref. [13].

Monthly averaged total solar irradiation on top surface area of pool's water is obtained using Equation 14-9 in GJ/month.

$$Q_{solor} = 0.0108 \times A \times \tau \times \alpha_{abs} \times I_{solor} \qquad 14\text{-}9$$

where I_{solar} is the monthly averaged total solar irradiation on top surface area of water in (kW-hr)/(m^2-day). I_{solar} is composed of direct beam radiation, diffuse radiation and ground reflected radiation components.

$I_{solar} = 5.3\ (kW - hr)/(m^2 - day)$ in the following calculations. τ is the hemispherical transmissivity averaged over wavelength of total solar radiation through top surface area of water and it has a value of 0.92 in the following calculations. α_{abs} is the hemispherical absorptivity for water averaged over wavelength of total solar radiation that reaches pool's surface and it is equal to 0.60 in the following calculations.

For T_w = 25 *C*, T_a = 15 *C*, *L* = 10 *m*, *W* = 8 *m* and *U* = 3.0 *m/s*, energy exchange between an uncovered pool's top water surface and the environment is calculated using Equations 14-1 through 14-9 and results are presented in Table 14-1. Similar results are presented for a UV stabilized transparent plastic covered pool for 12-hours at night in order to suppress heat losses due to evaporation are shown in Table 14-2. Also results are presented for a UV stabilized transparent bubble type covered pool for 12-hours at night in order to suppress heat losses both due to evaporation and convection are shown in Table 14-3.

Table 14-1 Energy transfer modes in GJ per month for an uncovered pool.

Energy Transfer Mode	Energy Transfer GJ/month	Percentage Of Total Energy Transfer
Radiation Loss	27.41	22.04
Convection Loss	14.81	11.90
Evaporation Loss	56.91	45.74
Solar Gain	25.28	20.32
Total Energy Transfer	124.41	100.00
Net Energy Loss	73.85	59.36

Table 14-2 Energy transfer modes in GJ per month for a UV stabilized transparent plastic covered pool for 12 hours every night in order to suppress heat losses due to evaporation.

Energy Transfer Mode	Energy Transfer GJ/month	Percentage Of Total Energy Transfer
Radiation Loss	27.41	28.57
Convection Loss	14.81	15.43
Evaporation Loss	28.45	29.66
Solar Gain	25.28	26.34
Total Energy Transfer	95.95	100.00
Net Energy Loss	45.40	47.31

Table 14-3 Energy transfer modes in GJ per month for a UV stabilized transparent bubble type covered pool for 12 hours every night in order to suppress heat losses both due to evaporation and convection.

Energy Transfer Mode	Energy Transfer GJ/month	Percentage Of Total Energy Transfer
Radiation Loss	27.41	30.96
Convection Loss	7.40	8.36
Evaporation Loss	28.45	32.13
Solar Gain	25.28	28.55
Total Energy Transfer	88.55	100.00
Net Energy Loss	30.59	34.55

If the pool water in the above example is being heated by natural gas at $3.00 per GJ and at 70% efficiency, an uncovered pool will expend $316.50 per month to keep up the 25 C pool water temperature. Same pool will expend for heating $194.56 per month ($121.94 saving per month), if the pool is covered by a UV stabilized transparent plastic for 12 hours every night. Same pool will expend for heating $131.10 per month ($185.41 saving per month), if the pool is covered by a UV stabilized transparent bubble type cover for 12 hours every night.

Above results are for a pool with a width of 8 m and a length of 10 m, namely a water top surface area of 80 m^2. If the length of the pool increases while keeping its width at 8 m, heat loss per month with respect to increasing pool length increases almost linearly as shown in Figure 14-1 for a pool with no cover, one with a plastic cover for 12 hours per day at night and one with bubble type cover for 12 hours per day at night. Sensitivity of heat loss to increasing pool length for a pool with no cover is shown in Figure 14-2. This sensitivity is about 6.77 GJ/(month-m).

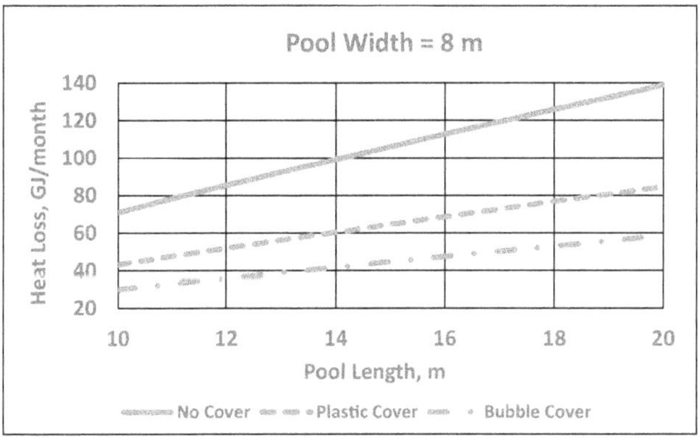

Figure 14-1 Heat loss per month versus pool length for a pool with no cover, one with a plastic cover for 12 hours per day at night and one with bubble type cover for 12 hours per day at night.

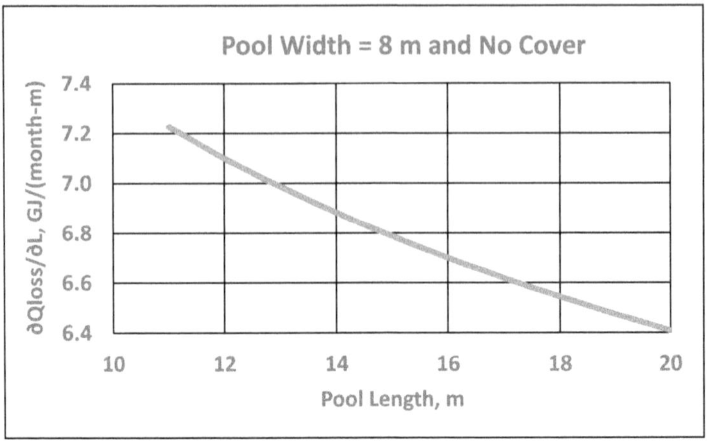

Figure 14-2 Sensitivity of heat loss per month to pool length versus pool length for an 8 meter wide pool with no cover.

If the wind velocity over the top surface of the pool increases while keeping its width at 8 m and its length at 10 m, heat loss per month with respect to increasing wind velocity is shown in Figure 14-3 for a pool with no cover, one with a plastic cover for 12 hours per day at night and one with bubble type cover for 12 hours per day at night. Pools with a bubble type cover suppresses best way evaporation and convection losses to air due to increasing wind velocity. Sensitivity of heat loss to increasing wind velocity for a pool with no cover is shown in Figure 14-4. This sensitivity on the average is about 6.02 (GJ-hr)/(month-km). Sensitivity of heat loss to increasing wind velocity decreases as the wind velocity increases.

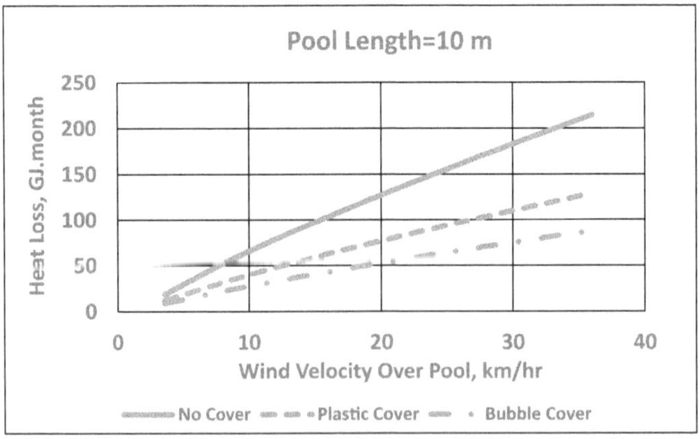

Figure 14-3 Heat loss per month versus wind velocity for a pool with no cover, one with a plastic cover for 12 hours per day at night and one with bubble type cover for 12 hours per day at night.

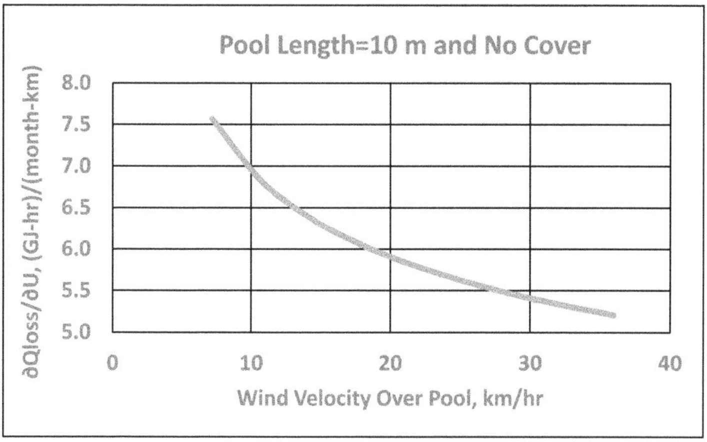

Figure 14-4 Sensitivity of heat loss per month to wind velocity over the top surface of the pool versus wind velocity for an 8 m by 10 m pool with no cover.

If the desired temperature for the pool water increases while keeping its width at 8 m and its length at 10 m, heat loss per month with respect to increasing water temperature is shown in Figure 14-5 for a constant air temperature of 16 C and for a pool with no cover, one with a plastic cover for 12 hours per day at night and one with bubble type cover for 12 hours per day at night. Sensitivity of heat loss to increasing water temperature for a pool with no cover is shown in Figure 14-6. This sensitivity on the average is about 7.02 GJ/(month-C). Sensitivity of heat loss to water temperature increases as the desired water temperature increases.

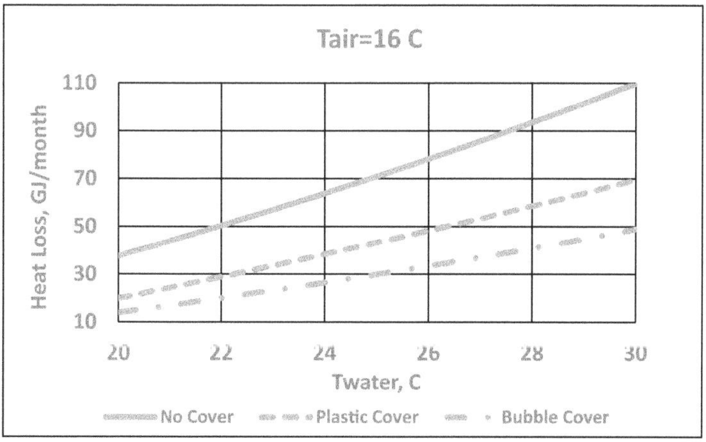

Figure 14-5 Heat loss per month versus water temperature for a pool with no cover, one with a plastic cover for 12 hours per day at night and one with bubble type cover for 12 hours per day at night.

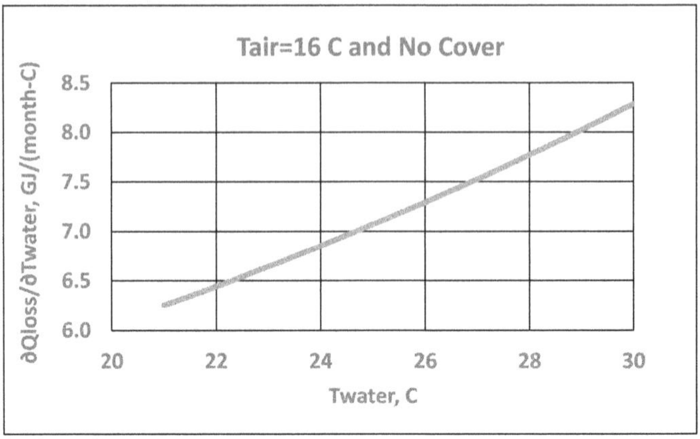

Figure 14-6 Sensitivity of heat loss per month to water temperature versus water temperature for an 8 m by 10 m pool with no cover.

If the outside air temperature over the pool increases while keeping its width at 8 m and its length at 10 m, heat loss per month with respect to increasing air temperature is shown in Figure 14-7 for a constant water temperature of 25 C and for a pool with no cover, one with a plastic cover for 12 hours per day at night and one with bubble type cover for 12 hours per day at night. Sensitivity of heat loss to increasing air temperature for a pool with no cover is shown in Figure 14-8. This sensitivity on the average is about - 3.28 GJ/(month-C). Absolute value of the sensitivity of heat loss to air temperature increases as the air temperature increases. When the air temperature is the same as the water temperature at 25 C, the convective heat transfer potential diminishes and there is no heat loss difference between the plastic cover and the bubble type cover.

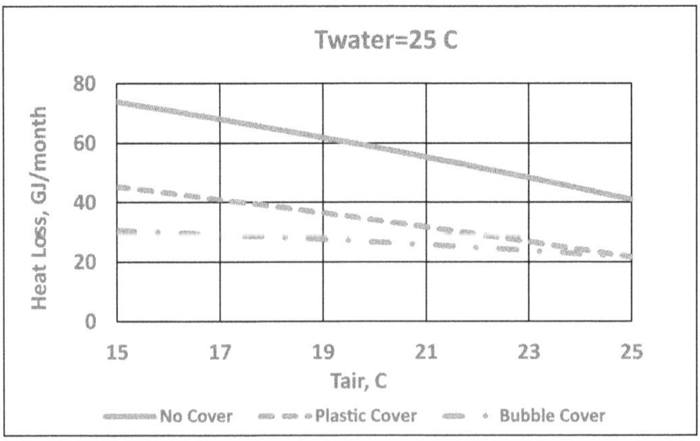

Figure 14-7 Heat loss per month versus air temperature for a pool with no cover, one with a plastic cover for 12 hours per day at night and one with bubble type cover for 12 hours per day at night.

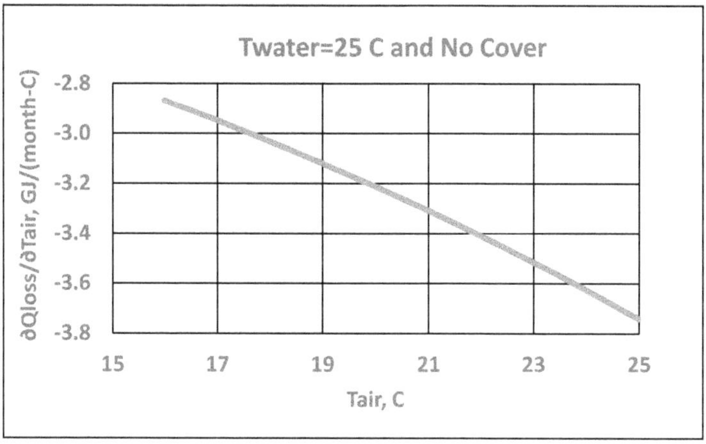

Figure 14-8 Sensitivity of heat loss per month to air temperature versus air temperature for an 8 m by 10 m pool with no cover.

15

Radiation Heat Transfer in Unsteady State

In this chapter, radiant heating of spherical balls with different radii will be analyzed. A spherical ball (1) will be placed inside at the center of a spherical oven (2) which has a temperature controlled inner surface for infrared radiant heating. The spherical ball is heated by radiation in a vacuum environment, namely there is no convection heat transfer at the surface of the spherical ball. See Figure 15-1.

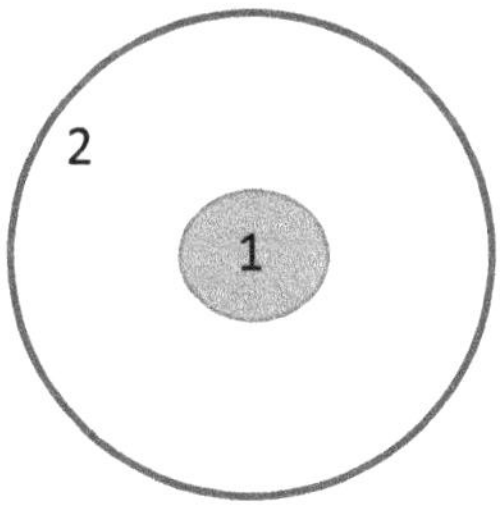

Figure 15-1 Unsteady radiant heating of a spherical ball at the center of a spherical oven in vacuum.

It is also assumed that the temperature inside the spherical ball is uniform at any instant of time during the unsteady radiant heating process, namely $Biot\ Number = \frac{h_r \times V_1}{A_1 \times k_1} < 0.1$. h_r is the radiation heat transfer coefficient in W/(m^2-K) defined by Equation 15-1.

$$h_r = \frac{\varepsilon_1 \times \sigma \times \left(T_2^4 - T_1^4\right)}{\left(T_2 - T_1\right)} \qquad 15\text{-}1$$

V_1 is the volume of the spherical ball in m^3. A_1 is the surface area of the spherical ball in m^2. k_1 is the thermal conductivity of the spherical ball material in W/(m-K). ε_1 is the hemispherical surface emissivity of spherical ball's averaged over all directions and wavelengths and it is equal to its surface absorptivity, namely assuming a gray surface. σ is the Stefan-Boltzmann constant at the value of 5.67x10^{-8} W/(m^2-K^4). T_1 is spherical ball's temperature in K. T_2 is the constant surface temperature of spherical oven's interior surface in K.

A characteristic length, L_1, is defined as $\frac{V_1}{A_1}$ which is $\frac{R_1}{3}$ for the spherical ball. Then the uniform interior temperature assumption made above for the spherical ball can be used if the internal resistance, $\frac{L_1}{k_1}$, is an order of magnitude less than the external radiant heat transfer resistance, $\frac{1}{h_r}$.

Infrared radiation heat transfer coming out from the interior surface of the spherical oven and reaching the exterior surface of the spherical ball, $Q_{2\to1}$, is defined in Equation 15-2.

$$Q_{2\to1} = A_2 \times F_{2\to1} \times \sigma \times \left(T_2^4 - T_1^4\right) \qquad \text{15-2}$$

Where $F_{2\to1}$ is the configuration factor that defines the fraction of radiant energy emitted diffusively and incident from the interior surface of the spherical oven to the exterior surface of the spherical ball which is located at the center of the oven. However, there is a reciprocity relationship between these two finite surfaces as shown in Equation 15-3.

$$A_2 \times F_{2\to1} = A_1 \times F_{1\to2} \qquad \text{15-3}$$

Also the configuration factor $F_{1\to2}$ = 1.0. Then Equation 15-2 can be rewritten as follows in Equation 15-4 for the radiant energy that is being absorbed by the spherical ball.

$$Q_{2\to1} = A_1 \times \varepsilon_1 \times \sigma \times \left(T_2^4 - T_1^4\right) \qquad \text{15-4}$$

Unsteady state of the spherical ball's temperature is defined in Equation 15-5 by the change in internal energy of the spherical ball being equal to the radiant energy that is being received from the interior surface of the spherical oven in a vacuum environment.

$$\rho_1 \times V_1 \times c_{p1} \times \frac{dT_1}{dt} = A_1 \times \varepsilon_1 \times \sigma \times \left(T_2^4 - T_1^4\right) \qquad \text{15-5}$$

where ρ_1 is the density of the spherical ball in kg/m^3 and c_{p1} is the specific heat at constant pressure of the spherical ball in J/(kg-K). Equation 15-5 is a first order nonlinear differential equation. It has an initial condition of $T_{1i} = 298\ K$ and it has an exact solution and also a numerical approximate solution as well. The exact solution is obtained by using the following integral Equation 15-6.

$$\int \frac{dx}{\left(a^4 - x^4\right)} = \left(\frac{1}{4 \times a^3}\right) \times \left[Ln\left|\frac{a+x}{a-x}\right| + 2 \times tan^{-1}\left(\frac{x}{a}\right)\right] \qquad 15\text{-}6$$

Using the integral in Equation 15-6, After separating the variables, Equation 15-5 takes the following form in Equation 15-7. Equation 15-7 assumes that all properties of the spherical ball such as k_1, p_1, c_{p1}, and ε_1 are constants and the spherical oven's surface temperature, T_2 is a constant during the unsteady radiant heating process.

$$\frac{(\varepsilon_1 \times \sigma)}{(\rho_1 \times L_1 \times c_{p1})} \times t = \left(\frac{1}{4 \times T_2^3}\right) \times \left[Ln\left|\frac{T_2 + T_1}{T_2 - T_1}\right| - Ln\left|\frac{T_2 + T_{1i}}{T_2 - T_{1i}}\right| + 2 \times tan^{-1}\left(\frac{T_1}{T_2}\right) - 2 \times tan^{-1}\left(\frac{T_{1i}}{T_2}\right)\right] \qquad 15\text{-}7$$

Equation 15-5 can also be solved by numerical approximation with small temperature steps, i.e $(T_{1(j)} - T_{1(j-1)}) = 10\ K$, as shown in Equations 15-8 and 15-9.

$$T_{1(j)} = T_{1(j-1)} + \left(\frac{h_{r(j-1)}}{\rho_1 \times L_1 \times c_{p1}}\right) \times \left(T_2 - T_{1(j-1)}\right) \times \left(t_{(j)} - t_{(j-1)}\right) \qquad 15\text{-}8$$

$$h_{r(j-1)} = \left(\varepsilon_1 \times \sigma\right) \times \frac{\left(T_2^4 - T_{1(j-1)}^4\right)}{\left(T_2 - T_{1(j-1)}\right)} \qquad 15\text{-}9$$

Equations 15-7 through 15-9 are solved for a spherical ball which is made out of a special steel with a radius of 0.2 m and at an initial temperature at 298 K and which is to be heated to 673 K with infrared radiation heating in a vacuum oven. The spherical steel ball is placed at the center of a spherical oven with a radius of 0.5 m and whose interior surfaces are kept at a constant temperature of 773 K for infrared heating. The spherical ball which is made out of a special steel which has the following properties.

$\varepsilon_1 = 0.4$, $\rho_1 = 7{,}800\ kg/m^3$, $k_1 = 40\ W/(m^2 - K)$

$c_{p1} = 550\ J/(kg - K)$

With above parameters, initially at t=0, $h_r = 16.67\ W/(m^2 - K)$. After reaching the desired 673 K temperature at t=4.84 hrs, $h_r = 34.45\ W/(m^2 - K)$.

With above h_r values, initially at t=0, the Biot # is 0.028. At t=4.84 hr, the Biot # is 0.057. So uniform interior temperature assumption for the spherical steel ball is a valid one. For the above case, time versus temperature plot for the exact solution and for the approximate numerical solution are shown in Figure 15-2. After 4.84 hours the spherical steel ball reached the desired temperature of 673 K using the exact solution. After 4.84 hours the spherical steel ball reached a higher temperature of 675.77 K, i.e. a 0.41% error, using the approximate numerical solution with $T_{1(j)} - T_{1(j-1)}) = 10\ K$ steps. More accurate numerical solutions can be achieved with smaller $(T_{1(j)} - T_{1(j-1)})$ steps.

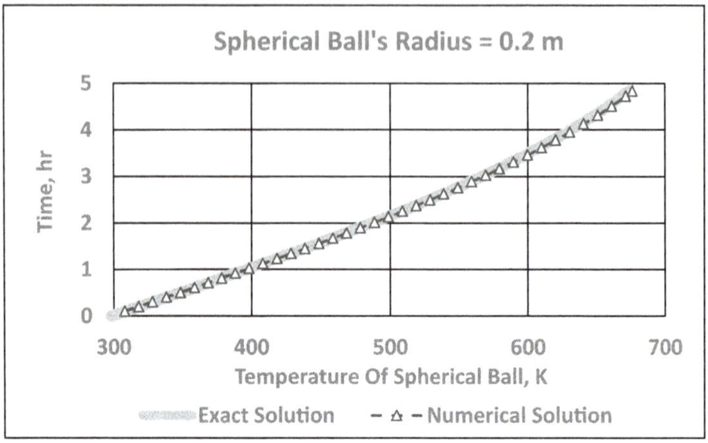

Figure 15-2 Time versus temperature profile for the spherical ball's temperature to reach 673 K by infrared heating. Comparing the exact solution with the numerical solution.

Time versus temperature profiles for the spherical ball's temperature to reach 673 K by infrared heating in a vacuum for different radii spherical balls that are made out of special steel with above given properties are shown in Figure 15-3. A spherical ball with 0.03 m radius reaches 673 K temperature in 0.73 hour. A spherical ball with 0.30 m radius reaches 673 K temperature in 7.27 hours.

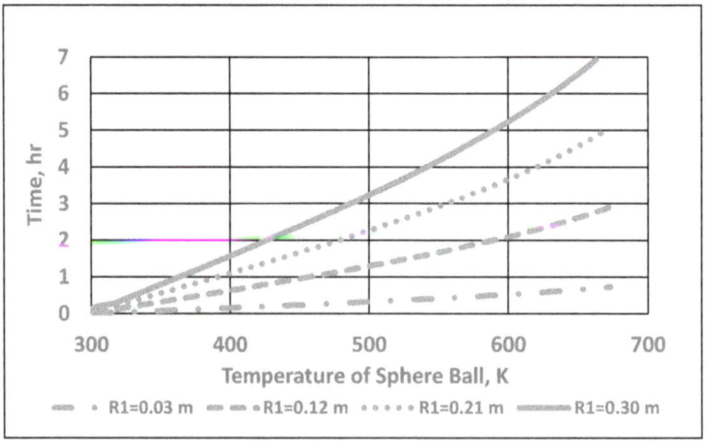

Figure 15-3 Time versus spherical ball's temperature for different radii spherical balls that are heated by infrared heating in a vacuum and are made out of special steel.

Sensitivity of time to sphere ball's radius versus temperature of the sphere ball is shown in Figure 15-4. This sensitivity of time to sphere ball's radius increases as the sphere ball's temperature increases. At a given sphere ball $\frac{\partial t}{\partial (R1)}$ temperature, is a constant which can be deduced from Equations 15-7 and 15-8.

The sensitivity of time to sphere ball's radius at 428 K is 6.7 hr/m. The sensitivity of time to sphere ball's radius increases to 24.2 hr/m at 673 K.

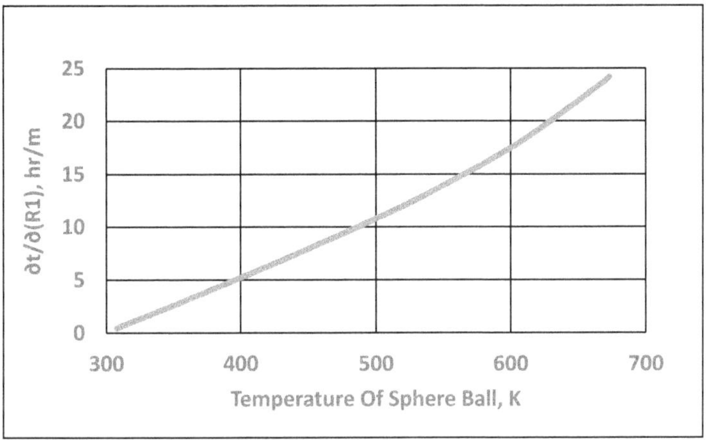

Figure 15-4 Sensitivity of time to sphere ball's radius versus temperature of the sphere ball.

At a given sphere ball temperature, $\frac{\partial t}{\partial \rho}$ is also a constant which can be deduced from Equations 15-7 and 15-8. The sensitivity of time to sphere ball's density at 673 K is 6.2×10^{-4} (hr-m^3)/kg.

At a given sphere ball temperature, $\frac{\partial t}{\partial c_p}$ is also a constant which can be deduced from Equations 15-7 and 15-8. The sensitivity of time to sphere ball's specific heat at constant pressure at 673 K is 8.8×10^{-3} (hr-kg-K)/J.

Times versus temperature for a spherical ball with a radius of 0.2 m and with different surface emissivities are shown in Figure 15-5. At a low surface emissivity of 0.2, it takes 9.69 hours to reach 673 K. As the surface emissivity goes up with a certain ratio, the time to reach a certain temperature goes down with the same ratio as expected. At a high surface emissivity of 0.4, it takes 4.84 hours to reach 673 K.

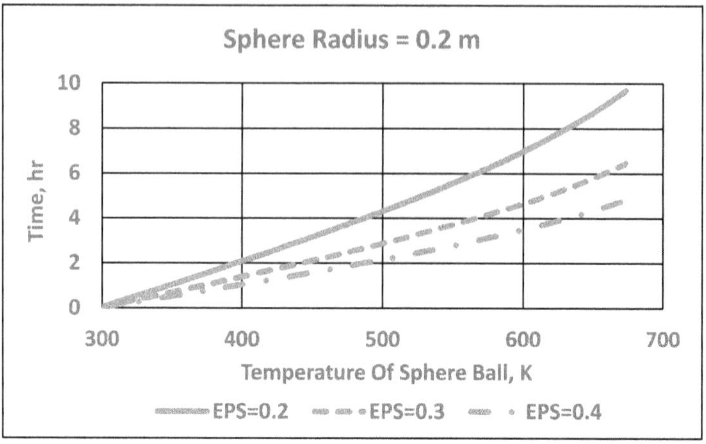

Figure 15-5 Times versus temperature for a spherical ball with a radius of 0.2 m and with different surface emissivities.

Sensitivities of time to surface emissivity versus surface emissivity for different radii spherical balls at 673 K are given in Figure 15-6. Absolute values of these sensitivities increase with increasing sphere ball's radius. Also absolute values of these sensitivities are more prominent at low surface emissivity values. As surface emissivities increase, absolute values of sensitivities decrease.

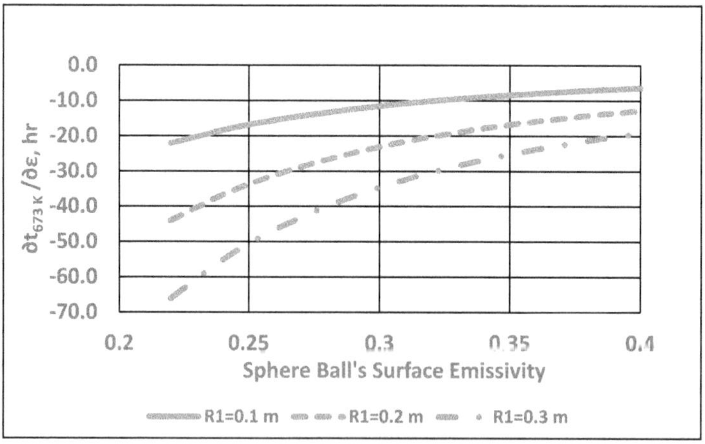

Figure 15-6 Sensitivities of time to surface emissivity versus surface emissivity for different radii spherical balls at 673 K.

16

Cooling of Steel Spherical Balls by Natural Convection and Radiation Heat Transfer in Unsteady State

In this chapter, steel spherical balls at different radii that were heated to 400 C (673 K) by infrared radiation in a vacuum environment in Chapter 15 will be cooled down slowly by natural convection and by infrared radiation heat transfer in unsteady state down to a temperature of 50 C (323 K). A hot spherical ball will be placed at the center of a large air chamber in which chamber air and chamber walls are kept at a constant temperature at 25 C (298 K). This kind of slow cooling down of hot metals are done in order to relieve stresses that are generated in the metal structures during their processing. It is also assumed that the temperature inside the spherical ball is uniform at any instant of time during the unsteady state cooling process, namely. *Biot number* $= \dfrac{h_t \times V_1}{A_1 \times k_1} < 0.1$. h_t is the total heat transfer coefficient which is the sum of the radiation, h_r, and the natural convection, h_{nc}, heat transfer coefficients in W/(m²-K) as defined by Equations 16-1 to 16-3.

$$h_t = h_r + h_{nc} \qquad \text{16-1}$$

$$h_t = \frac{\varepsilon_1 \times \sigma \times \left(T_1^4 - T_e^4\right)}{\left(T_1 - T_e\right)} \qquad \text{16-2}$$

$$h_{nc} = \left(\frac{Nu_D \times k}{D}\right) \qquad \text{16-3}$$

where the average Nusselt number for the whole surface of a sphere, Nu_D, in Equation 16-3 is given in Equation 16-4. For this empirical correlation Equation 16-4, see Ref. [13]. Nu_D represents the average temperature gradient at the entire surface of the sphere due to natural convection heat transfer. Subscript (1) refers to the spherical ball and subscript (e) refers to the air chamber environment.

$$Nu_D = 2 + \frac{0.589 \times Ra_D^{0.25}}{\left[1 + (0.469 / Pr)^{9/16}\right]^{4/9}} \qquad 16\text{-}4$$

Ra_D is the dimensionless Rayleigh number which is the product of Grashof number, $Gr_D = \frac{g \times \beta \times (T_1 - T_e) \times D^3}{\nu^2}$, which is the ratio of buoyancy forces to viscous forces acting on the air elements around the boundary layer of the sphere, and the Prandtl number, $Pr = \frac{\nu}{\alpha}$, which is the ratio of momentum and thermal diffusivities for air in the boundary layer of the sphere. The Rayleigh number is given in Equation 16-5.

$$Ra_D = \frac{g \times \beta \times (T_1 - T_e) \times D^3}{\nu \times \alpha} \qquad 16\text{-}5$$

The empirical correlation given in Equation 16-4 for the Nusselt number is valid for $Ra_D < 1 \times 10^{11}$ and for $Pr > 0.7$. k is the thermal conductivity of air in W/(m-K). g is the gravitational constant at 9.8 m/s^2. β is the volumetric thermal expansion coefficient for air in 1/K. ν is the kinematic viscosity of air in m/s^2. α is the thermal diffusivity of air in m/s^2. During following calculations, all air properties in the boundary layer of the sphere are taken at the average film temperature of 398 K. This film temperature is the average between the initial sphere ball temperature at 400 C (673 K) and the temperature when the sphere balls are taken out of the cooling chamber as they reach to 50 C (323 K) and the constant air chamber temperature at 25 C (298 K).

The unsteady state of a sphere cooling, namely change in internal energy of the spherical ball with respect to time, t, in seconds in a large chamber in which chamber air and chamber walls are kept at a constant temperature, can be defined by the following energy Equation 16-6.

$$\rho_1 \times V_1 \times c_{p1} \times \frac{dT_1}{dt} = -A_1 \times h_t \times (T_1 - T_e) \qquad 16\text{-}6$$

ρ_1 is the density of the spherical ball in kg/m^3. V_1 is the volume of the spherical ball in m^3. c_{p1} is the specific heat at constant pressure of the spherical ball in J/(kg-K). A_1 is the surface area of the spherical ball in m^2. k_1 is the thermal conductivity of the spherical ball material in W/(m-K). ε_1 is the hemispherical surface emissivity of spherical ball's averaged over all directions and wavelengths and it is equal to its surface absorptivity, namely assuming a gray surface. σ is the Stefan-Boltzmann constant at the value of 5.67×10^{-8} W/(m^2-K^4). T_1 is spherical ball's time dependent temperature in K. T_e is the constant temperature of the cooling chamber's air and walls in K.

A characteristic length, L_1, is defined as $\frac{V_1}{A_1}$ which is $\frac{R_1}{3}$ for the spherical ball. Then the uniform interior temperature assumption made above for the spherical ball can be used if the internal resistance, $\frac{L_1}{k_1}$, is an order of magnitude less than the total heat transfer resistance, $\frac{1}{h_t}$, i.e. *Biot number* < 0.1.

Equations 16-1 through 16-6 constitute a first order non-linear differential equations. These equations are solved numerically using the following approximate Equations 16-7 to 16-10. The initial condition at $t = 0$ is $T_1 = 673$ *K*. Also a time interval of 0.25 hr is used with very accurate results for the present numerical calculations, namely $(t_{(j)} - t_{(J-1)}) = 900$ *s* in Equation 16-7.

$$T_{1(j)} = T_{1(j-1)} - \left(\frac{h_{t(j-1)}}{\rho_1 \times L_1 \times c_{p1}}\right)\times\left(T_{1(j-1)} - T_e\right)\times\left(t_{(j)} - t_{(j-1)}\right) \qquad 16\text{-}7$$

$$h_{t(j-1)} = h_{r(j-1)} + h_{nc(j-1)} \qquad 16\text{-}8$$

$$h_{r(j-1)} = \frac{\varepsilon_1 \times \sigma \times \left(T_{1(j-1)}^4 - T_e^4\right)}{\left(T_{1(j-1)} - T_e\right)} \qquad 16\text{-}9$$

$$h_{nc(j-1)} = \left(\frac{Nu_{D(j-1)} \times k}{D}\right) \text{ where } Ra_{D(j-1)} = \frac{g \times \beta \times \left(T_{1(j-1)} - T_e\right)\times D^3}{\nu \times \alpha} \qquad 16\text{-}10$$

The uniform interior temperature assumption made for the spherical ball is valid in all of the calculations, since the maximum Biot number is 0.044 for the largest sphere ball with a radius of 0.3 m and with a surface emissivity of 0.40 when the maximum total heat transfer coefficient is 18.17 W/(m^2 - K) at t=0.

Following property values are used in the numerical calculations of Equations 16-7 to 16-10.

Spherical ball properties: $R_1 = 0.03$ *to* 0.30 *m*, $\varepsilon_1 = 0.20$ *to* 0.40, $k_1 = 40$ *W/(m^2 – K)*, $\rho_1 = 7800$ *kg/m^3*, $c_{p1} = 550$ *K/(kg – K)*

Air properties at the average film temperature of 398 *K* :
$\nu = 2.64 \times 10^{-5}$ *m^2/s*, $a = 3.80 \times 10^{-5}$ *m^2/s*, *Pr* = 0.70,

$$\beta \cong \frac{1}{T_{e-film}} = 0.00251\ \text{K}^{-1}.$$

Sphere ball's temperatures versus cooling times for different radii and with an emissivity of 0.4 are shown in Figure 16-1. As the sphere ball's radius gets smaller, it reaches the final temperature of 50 C faster as expected. A steel spherical ball with a radius of 0.03 m cools down to 50 C (323 K) in 2.12 hours. However, a spherical ball with a radius of 0.30 m, namely with a 1,000 times larger volume, cools down to 50 C (323 K) in 39.42 hours, i.e. 18.6 times slower.

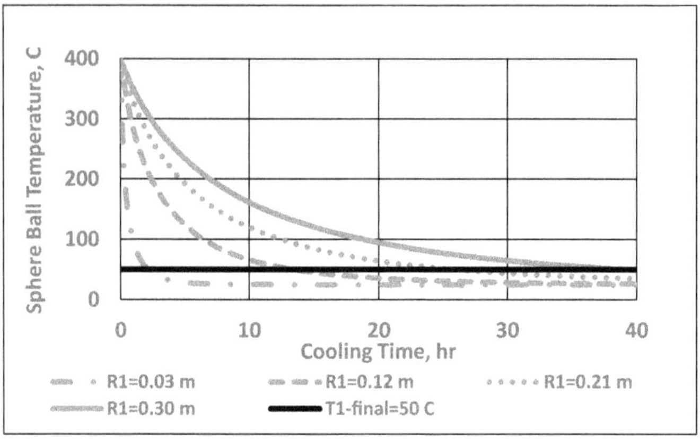

Figure 16-1 Sphere ball's temperatures versus cooling times for different radii.

Sensitivity of time to cool from 400 C (673 K) to 50 C (323 K) to sphere ball's radius versus sphere's ball radius is shown in Figure 16-2. At small radius balls, i.e. R1=0.03 m, this sensitivity is 117 hr/m. As the ball radius increases so does the sensitivity to cool down to 50 C (323 K). At large radius balls, i.e. R1=0.30 m, this sensitivity increases to 149 hr/m.

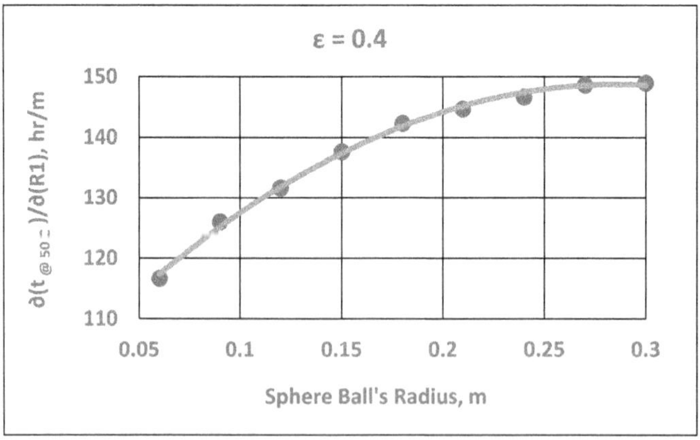

Figure 16-2 Sensitivity of time to cool from 400 C (673 K) to 50 C (323 K) to sphere ball's radius versus sphere's ball radius.

As the cooling time increases and ball temperature approaches the environment temperature, radiation heat transfer coefficient in Equation 16-9 decreases as shown in Figure 16-3. Small radius balls with small characteristic lengths accelerates the temperature drop with respect to time, see Equation 16-7, and therefore decreases the radiation heat transfer coefficient faster. At R1=0.03 m, h_r value starts at a value of 11.93 W/(m^2-K) at t=0 and at T1=400 C (673 K) and reduces to a value of 2.73 W/(m^2-K) at t=2.12 hours and at T1=50 C (323 K) As the spherical ball's radius gets larger, the temperature drop slows down with respect to time and therefore decreases the radiation heat transfer coefficient slower. At R1=0.30 m, h_r value again starts at a value of 11.93 W/(m^2-K) at t=0 and at T1=400 C (673 K) and reduces to a value of 2.72 W/(m^2-K) at t=39.42 hours and at T1=50 C (323 K). At large cooling hours, i.e. $t > 40$ *hours*, h_r asymptotes to a value of 2.41 W/(m^2-K) for all spherical ball's radii.

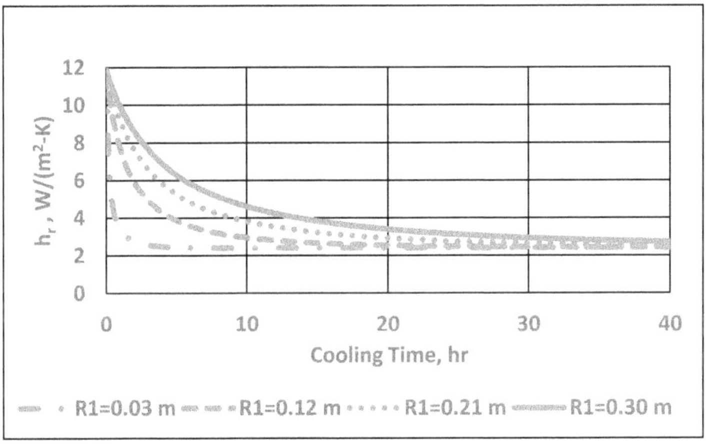

Figure 16-3 Radiation heat transfer coefficients versus cooling time for different radii spherical balls.

As the cooling time increases and ball temperature approaches the environment temperature, natural convection heat transfer coefficient in Equation 16-10 decreases as shown in Figure 16-4. Small radius balls with small characteristic lengths accelerates the temperature drop with respect to time, see Equation 16-7, and therefore decreases the natural convection heat transfer coefficient faster. At R1=0.03 m, h_{nc} value starts at a value of 10.88 W/(m^2-K) at t=0 and at T1=400 C (673 K) and reduces to a value of 6.0 W/(m^2-K) at t=2.12 hours and at T1=50 C (323 K). As the spherical ball's radius gets larger, the temperature drop slows down with respect to time and therefore decreases the natural convection heat transfer coefficient. At R1=0.30 m, h_{nc} value starts at a value of 5.60 W/(m^2-K) at t=0 at T1=400 C (673 K) and reduces to a value of 2.85 W/(m^2-K) at t=39.42 hours and at T1=50 C (323 K).

For small radii sphere balls, i.e. R1=0.03 m, Nu_D approaches the value of 1.0 W/(m^2-K) fast, namely the natural convection heat transfer resistance approaches the air conduction resistance value at the boundary layer around the sphere. As the sphere's radius increases, the natural convection heat transfer coefficient reduces more slowly due to slower cooling down.

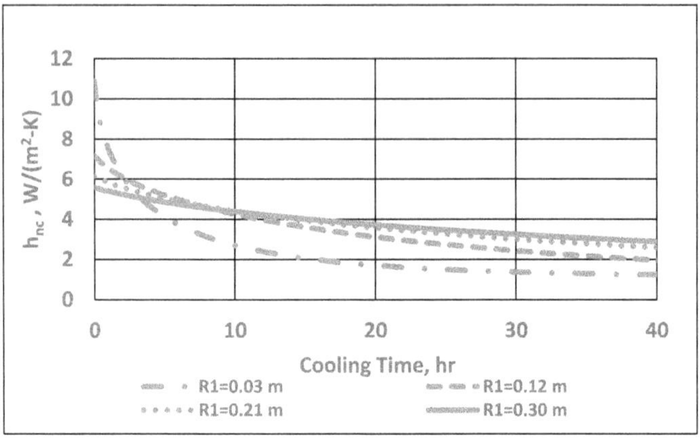

Figure 16-4 Natural convection heat transfer coefficients versus cooling time for different radii spherical balls.

Total heat transfer coefficients versus cooling time for different radii spherical balls are shown in Figure 16-5. h_t starts at a value of 22.81 W/(m^2-K) and its initial value decreases to 17.53 W/(m^2-K) as the sphere ball's radius increases from 0.03 to 0.30 m. h_t asymptotes to a value of 3.60 W/(m^2-K) at large cooling times, i.e. $t > 40$ *hours* for R1=0.03 m. The asymptote value of h_t increases as the sphere ball's radius increases. At R1=0.30 m, the asymptote value of h_t is 5.45 W/(m^2-K).

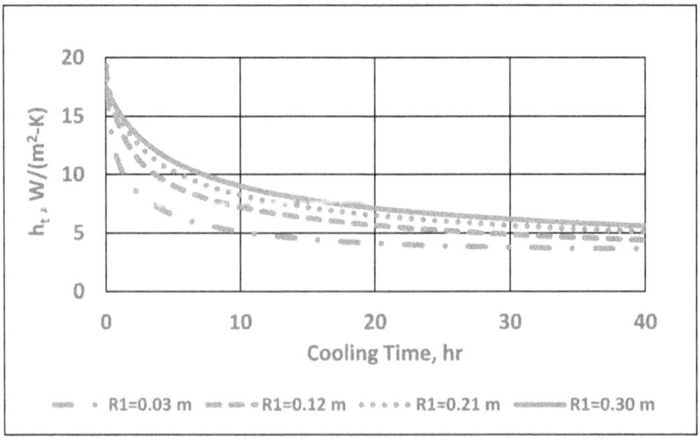

Figure 16-5 Total heat transfer coefficients versus cooling time for different radii spherical balls.

Cooling time of a spherical steel ball is also affected by its surface emissivity characteristics. As the surface emissivity increases linearly so does the h_r. Spherical balls temperatures versus cooling times for two different emissivities, i.e. $\varepsilon = 0.2$ *and* 0.4, and for three different radii, i.e. $R_1 = 0.03, 0.15$ *and* $0.30\ m$, are shown in Figure 16-6. As the emissivity increases for a given radius spherical ball, cooling time to a desired temperature decreases. Changing emissivity effect on cooling times diminish as sphere's radius gets small. See the curves for $R_1 = 0.03\ m$ in Figure 16-6.

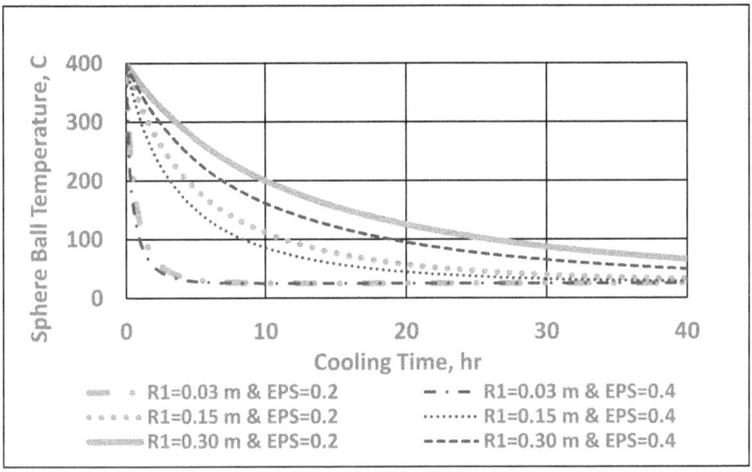

Figure 16-6 Spherical balls temperatures versus cooling times for two different emissivities and for three different radii.

Times to cool down from 400 C (673 K) to 50 C (323 K) versus spherical ball's surface emissivity for two different radii balls are shown in Figure 16-7. Cooling time for a 0.30 m radius ball is 53 hours with a surface emissivity of 0.2. This cooling time reduces by 27.5% to 39.5 hours when the surface emissivity increases to 0.4. Cooling time for a 0.15 m radius ball is 23 hours with a surface emissivity of 0.2. This cooling time reduces by 22.8% to 17.8 hours when the surface emissivity increases to 0.4.

Sensitivity of cooling time to 50 C (323 K) to emissivity, i.e., $\frac{\partial t_{T1-50c}}{\partial \varepsilon}$ decreases as emissivity increases. This decreasing sensitivity with increasing emissivity also reduces as the steel sphere's radius reduces. For example for a 0.3 m radius ball, $\frac{\partial t_{T1-50c}}{\partial \varepsilon} = -84.4\ hr$ at a low emissivity of 0.24 and it decreases to, $\frac{\partial t_{T1-50c}}{\partial \varepsilon} = -50.0\ hr$ at a high emissivity of 0.4. However, for a 0.15 m radius ball, $\frac{\partial t_{T1-50c}}{\partial \varepsilon} = -31.3\ hr$ at a low emissivity of 0.24 and it decreases to, $\frac{\partial t_{T1-50c}}{\partial \varepsilon} = -18.8\ hr$ at a high emissivity of 0.4.

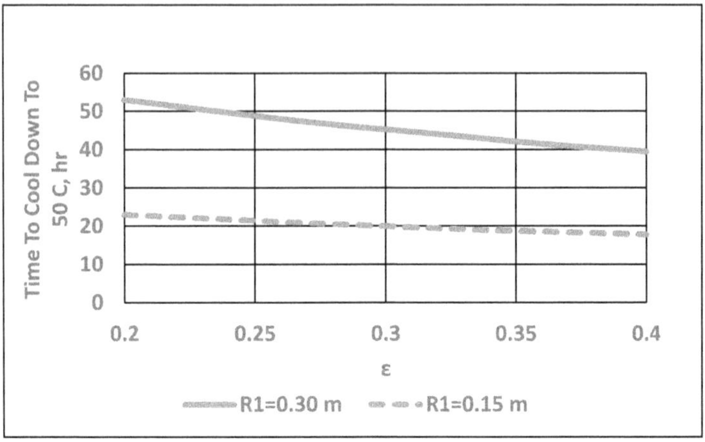

Figure 16-7 Time to cool down from 400 C (673 K) to 50 C (323 K) versus spherical ball's surface emissivity for two different radii balls.

17

Baking Pizza by Infrared Radiation and Conduction Heat Transfers and Mass Transfer in Unsteady State

In this chapter, we will investigate baking of a pizza by infrared radiation heat transfer from hot ceiling bricks of a hemispherical dome of a wood fired oven and by conduction heat transfer from hot bricks covering the floor which are contacting the bottom surface of a pizza crust. During the baking process, pizza will also lose heat by mass transfer, namely by evaporation, from pizza's top surface to oven's air. All heat transfer mechanisms acting on the pizza are in unsteady state from the moment the pizza is placed on the floor of the oven. Objects of baking a good pizza is not to burn the bottom of the crust and have a crispy bottom and also cook the dough and toppings to an exact water content at the end of baking process. We will assume the pizza to be a large disk, i.e. a diameter of D = 0.25 m, with a small thickness, i.e. 0.01 m, and use the one dimensional unsteady state heat transfer equation presented in Equation 17-1.

$$\rho_p \times V \times c_{pp} \times \frac{dT}{dt} = Q_{rad} + Q_{cond} - Q_{evap} \qquad 17\text{-}1$$

where ρ_p is the weight averaged density of pizza at 940 kg/m^3 for 8 mm thick crust and 2 mm thick toppings. V is the total volume of the pizza at 4.91×10^{-4} m^3 of which 3.93×10^{-3} m^3 is the volume of the crust and 4.91×10^{-5} m^3 is the volume of toppings. c_{pp} is the weight averaged specific heat at constant pressure of pizza at 0.424 J/(kg-K). T is pizza's time dependent top surface temperature in K which is initially assumed to be at 25 C (298 K). t is the pizza baking time in seconds. Q_{rad} is the infrared radiation heat transfer from ceiling bricks of the oven in Watts to pizza's top surface and it is given in Equation 17-2.

$$Q_{rad} = A \times \varepsilon \times \sigma \times \left(T_c^4 - T^4\right) \quad \text{17-2}$$

where A is the surface area of the top of pizza and it is 0.0569 m^2. ε is the hemispherical surface emissivity of pizza's top surface averaged over all directions and wavelengths and it is equal to its surface absorptivity, namely assuming a gray surface. The hemispherical surface emissivity of pizza's top surface is assumed to be very close to being a black surface, namely has a value of unity. α is the Stefan-Boltzmann constant at the value of 5.67×10^{-8} W/(m^2-K^4). T_c is the constant temperature of oven's ceiling bricks in K.

Q_{cond} is the transient conduction heat transfer from oven's floor bricks to the bottom of pizza's crust in Watts and it is given in Equation 17-3. In Equation 17-3, it is assumed that oven's floor bricks and the bottom of the pizza's crust are semi-infinite bodies at different temperatures and they come into interfacial contact at $t = 0$ with negligible contact resistance. Also a constant interfacial temperature between these two bodies are assumed during the one second time interval that is used for numerical calculations. See Ref. [13].

$$Q_{cond} = A_b \times \left(T_f - T_{int}\right) \times \frac{k_b}{\left(\pi \times \alpha_b \times t\right)} \quad \text{17-3}$$

where A_b is pizza's bottom surface area in m^2. T_j is the constant temperature of bricks covering oven's floor in K where the pizza is placed for baking. k_b is the thermal conductivity of floor bricks in W/(m-K) at constant floor brick temperature. α_b is the thermal diffusivity of floor bricks in m^2/s at constant floor brick temperature. T_{int} is the constant interfacial crust contact temperature between the floor bricks and the bottom of the pizza crust in K during a one second time intervals. T_{int} is given in Equation 17-4. See Ref. [13].

$$T_{int} = \frac{T_f + \gamma \times T}{1 + \gamma} \quad \text{17-4}$$

where $\gamma = \sqrt{\frac{k_p \times \rho_p \times c_{pp}}{k_b \times \rho_b \times c_{pb}}}$. The subscript p stands for the weight averaged pizza properties at the average film temperature between T_f and T. The subscript b stands for the floor brick properties at the average film temperature between T_f and T.

Q_{evap} is the mass transfer loss due to evaporation from the top surface of the pizza to oven's air in Watts and it is given by Equation 17-5.

$$Q_{evap} = M \times H_{fg} \quad \text{17-5}$$

where M is the evaporation rate in kg/(m^2-hr) and H_{fg} is the latent heat of evaporation of water in air in (W-hr)/kg. The evaporation rate of water from the top surface of the pizza can be defined in terms of a convection mass transfer coefficient similar to the convection heat transfer coefficient and also by assuming the water vapor to be a perfect gas is shown in Equation 17-6.

$$M = h_m \times \frac{(p_w - RH \times p_a)}{R \times T} \quad 17\text{-}6$$

where h_m is the mass transfer coefficient at the boundary layer between pizza's top surface and the oven air in m/hr. p_w is the saturation pressure of water vapor in pizza's top surface in N/m^2. p_a is the saturation pressure of water vapor in oven's air in N/m^2. RH is the average relative humidity for a very dry oven air. R is the gas constant for water vapor with the value of 461.9 (N-m)/(kg-K). Saturation pressures for water vapor in Pascal are obtained for a particular surrounding temperature T_{surr} in C from Equation 17-7 (August-Roche-Magnus formula) which comes from a best fit empirical equation to experimental data.

$$p_{sat} = 610.94 \times exp\left(\frac{17.62 \times T_{surr}}{T_{surr} + 243.04}\right) \quad 17\text{-}7$$

The mass transfer coefficient can be obtained by analogy from natural convection heat transfer coefficient for the top surface of a heated plate. See Ref. [13]. These two processes of mass and heat transfer are analogous as shown in Equation 17-8.

$$h_m = \frac{h_{nat}}{\rho_{air} \times c_{p-air} \times Le^{0.667}} \quad 17\text{-}8$$

where $Le = \frac{\alpha_{air}}{D}$ is the Lewis number, the ratio of thermal diffusivity of oven air to diffusion coefficient of water in oven air. Lewis number is close to unity in our calculations. h_{nat} is the average natural convection heat transfer coefficient obtained from Equations 17-9 for the top surface of a heated plate.

$$Nu_L = \frac{h_{nat} \times L}{k_{air}} = 0.54 \times Ra_L^{0.25} \; for\, 10^4 \le Ra_L \le 10^7 \quad 17\text{-}9a$$

$$Nu_L = \frac{h_{nat} \times L}{k_{air}} = 0.54 \times Ra_L^{0.333} \; for\; 10^7 \le Ra_L \le 10^{11} \quad 17\text{-}9b$$

Ra_L is the dimensionless Rayleigh number which is the product of Grashof number, $Gr_L = \frac{g \times \beta \times (T_1 - T_e) \times L^3}{\nu^2}$, which is the ratio of buoyancy forces to viscous forces acting on the air elements around the boundary layer at pizza's top surface, and the Prandtl number, $Pr = \frac{\nu}{\alpha}$, which is the ratio of momentum and thermal diffusivities for oven air in the boundary layer at pizza's top surface. The Rayleigh number is given in Equation 17-10.

$$Ra_L = \frac{g \times \beta \times (T_1 - T_e) \times L^3}{\nu \times \alpha} \qquad 17\text{-}10$$

where g is the gravitational constant at 9.8 m/s^2. β is the volumetric thermal expansion coefficient for oven air in 1/K. ν is the kinematic viscosity of oven air in m/s^2. α_{air} is the thermal diffusivity of oven air in m/s^2. During following calculations, all oven air properties in the boundary layer at pizza's top surface are taken at an average film temperature. L is the characteristic length for pizza's top surface which is the top surface area divided by top surface's perimeter. L = 0.0625 m in the following calculations.

Properties of pizza during the process such as the diffusion coefficients are obtained from Ref. [7]. Equations 17-1 through 17-10 are solved by numerical methods using the explicit finite difference method using Equation 17-11 with one second time intervals.

$$T_{(j)} = T_{(j-1)} + \left(\frac{t_{(j)} - t_{(j-1)}}{\rho_p \times V \times c_{pp}} \right) \times \left(Q_{rad(j-1)} + Q_{cond(j-1)} - Q_{evap(j-1)} \right) \qquad 17\text{-}11$$

It should be emphasized that the interfacial crust contact temperature, T_{int}, between the floor bricks and the bottom of the pizza is assumed to be a constant during the one second time intervals of our numerical calculations.

In the present calculations, following two criteria are used to determine conditions for a well baked pizza. These two important criteria come from centuries of pizza baking experiments that resulted in a crispy crust with no burns and in well-cooked toppings.

1. Bottom surface of the pizza crust is crispy (not burned) when the final interfacial crust contact temperature is between 200 C and 210 C.
2. Pizza as a whole is well baked when its crust plus its toppings lose 20% of their water content.

Pizza's top surface temperatures versus baking time for oven ceiling bricks at 350 C and for three different oven floor bricks' temperatures at 200 C, 250 C and 300 C are shown in Figure 17-1. As pizza's top surface temperatures increase during the first 50 seconds of baking, it goes over the boiling temperature of water into superheated region depending upon the oven floor bricks' temperature.

During the baking process, infrared radiation heat transfer onto pizza's top surface pretty much stays a constant. Also as pizza's top surface temperature increases, conduction heat transfer from the bottom of the pizza decreases fast. Decreasing conduction heat transfer suppresses the heat loss due to mass transfer from pizza's top surface to oven air. Top surface temperature asymptotes at around 100 C. At low floor bricks' temperature of 200 C, top surface temperature approaches 100 C at about 100 seconds and stays around 100 C for rest of the baking process.

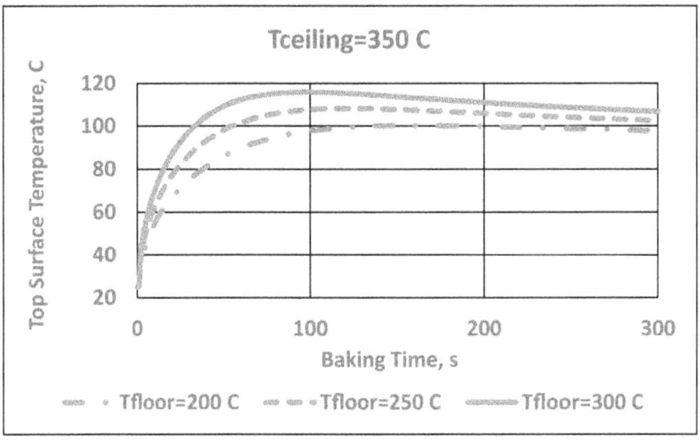

Figure 17-1 Pizza's top surface temperature versus baking time for oven ceiling bricks at 350 C and for three different oven floor bricks' temperatures.

Three heat transfer modes plus the net heat transfer curve versus baking time for oven ceiling bricks at 350 C and oven floor bricks' temperatures at 200 C are shown in Figure 17-2. Curves go up to pizza's 20% weight loss baking time at 276 seconds. Q_{net} falls to zero after 150 seconds of baking due to decreasing conduction heat transfer from the bottom of the pizza. Three heat transfer modes plus the net heat transfer curve versus baking time for oven ceiling bricks at 350 C and oven floor bricks' temperatures at 250 C are shown in Figure 17-3. Curves go up to pizza's 20% weight loss baking time at 217 seconds. Q_{net} falls to zero after 120 seconds of baking due to decreasing conduction heat transfer from the bottom of the pizza. Three heat transfer modes plus the net heat transfer curve versus baking time for oven ceiling bricks at 350 C and oven floor bricks' temperatures at 300 C are shown in Figure 17-4. Curves go up to pizza's 20% weight loss baking time at 175 seconds. Q_{net} falls to zero after 100 seconds of baking due to decreasing conduction heat transfer from the bottom of the pizza.

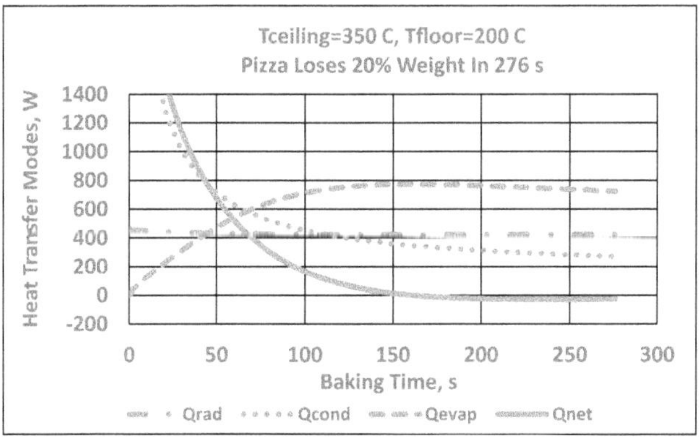

Figure 17-2 Radiation, conduction and evaporation heat transfers versus baking time for $T_{ceiling}$=350 C and T_{floor}=200 C. Pizza has to come out of the oven in 276 seconds.

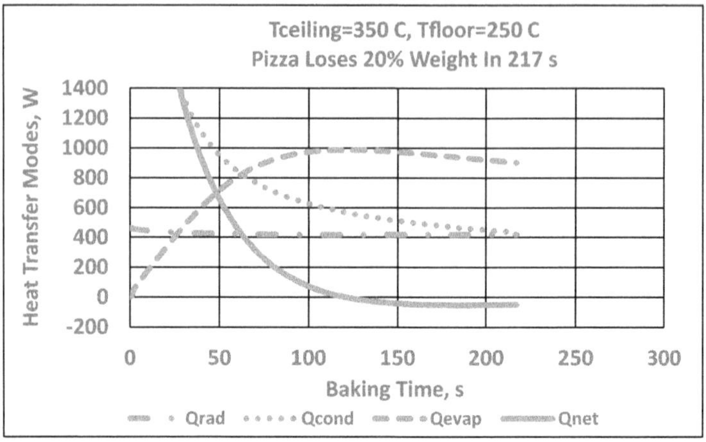

Figure 17-3 Radiation, conduction and evaporation heat transfers versus baking time for $T_{ceiling}$=350 C and T_{floor}=250 C. Pizza has to come out of the oven in 217 seconds.

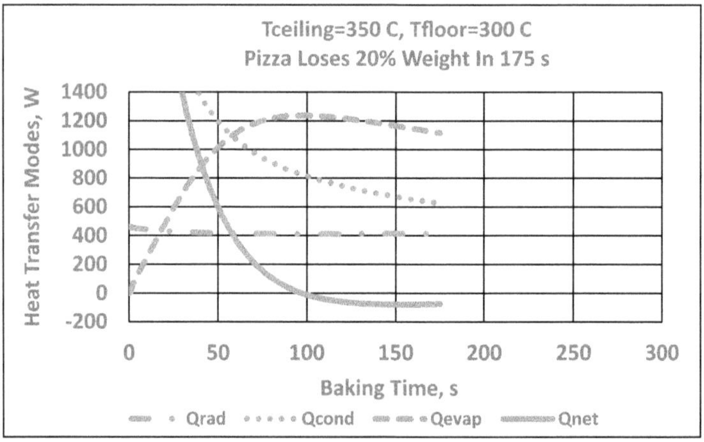

Figure 17-4 Radiation, conduction and evaporation heat transfers versus baking time for $T_{ceiling}$=350 C and T_{floor}=300 C. Pizza has to come out of the oven in 175 seconds.

Figure 17-5 shows interfacial crust contact temperatures, T_{int}, between the floor bricks and the bottom of the pizza versus baking time for $T_{ceiling}$=350 C and for six different floor bricks' temperatures at 250 C, 260 C, 270 C, 280 C, 290 C and 300 C. For an oven ceiling bricks' temperature of 350 C, for a good crispy crust without burning it, requires an oven floor bricks' temperature between 270 C and 280 C, namely an interfacial crust contact temperature between 200 C and 210 C.

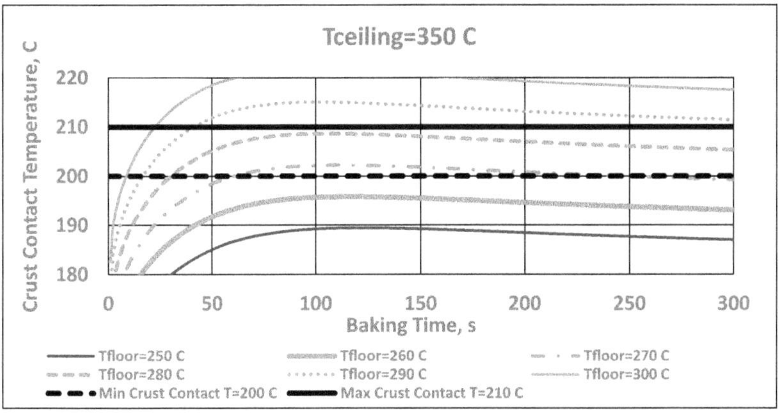

Figure 17-5 Interfacial crust contact temperature, T_{int}, between the floor bricks and the bottom of the pizza versus baking time for $T_{ceiling}$=350 C and for six different floor bricks' temperatures.

Figure 17-6 shows interfacial crust contact temperatures, T_{int}, between the floor bricks and the bottom of the pizza versus oven floor bricks' temperature for three different ceiling bricks' temperatures at 300 C, 350 C and 400 C. Some pizza makers increase the oven ceiling bricks' temperature in order to increase their throughput. Then they have to decrease their oven floor bricks' temperature so that pizza crust bottom does not burn.

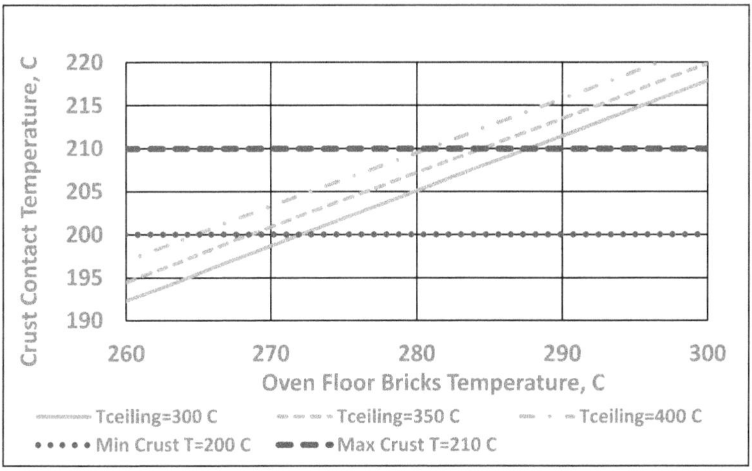

Figure 17-6 Interfacial crust contact temperature, T_{int}, between the floor bricks and the bottom of the pizza versus oven floor bricks' temperature for three different ceiling bricks' temperatures.

An oven's ceiling and floor bricks' temperatures have to be set up in a narrow range in order to achieve a baking process without burning the crispy crust. Feasible oven floor bricks' temperature region of operation for different oven ceiling bricks' temperatures are presented in Table 17-1.

Table 17-1 Maximum floor bricks' temperatures to achieve 210 C and 200 C interfacial crust contact temperatu re, T_{int}, between the floor bricks and the bottom of the pizza for 300 C, 350 C and 400 C ceiling bricks' temperatures.

	Ceiling Bricks' Temperature, C	Ceiling Bricks' Temperature, C	Ceiling Bricks' Temperature, C
	300	350	400
Max. Floor Bricks' Temp. In C To Achieve 210 C Contact Temp.	287.7	284.4	280.7
Min. Floor Bricks' Temp. In C To Achieve 200 C Contact Temp.	272.0	268.6	264.8

18

Microwave Radiation in Cooking Ovens

In this chapter, we will investigate microwave radiation in cooking ovens. Microwaves, in the electromagnetic spectrum, fall between the infrared radiation at the high frequency end, i.e. 300 GHz at 1 mm wavelength, radio waves at the low frequency end, i.e. 300 MHz at 1 m wavelength. Different bands of microwaves are used in different technologies ranging from garage door openers, point-to-point wireless communication links, satellite and spacecraft communications to advanced weather forecasting by providing global view of weather patterns and surface temperatures. Cosmic microwave background radiation fills the entire universe and it provides us knowledge about the beginnings of our universe, namely about the Big Bang. theory.

In a microwave oven, electromagnetic waves generated by a magnetron at the frequency of 2.45 GHz, a wavelength of 12.2 cm, enter an oven's cavity and penetrate the food which is placed at the center of a rotating turntable. Microwaves pass through most food containers made out of glass, paper and plastic without heating them. Part of the food molecules such as water or fat carry a slight positive charge while the other part of these food molecules carry a slight negative charge. This charge imbalance in food molecules colliding with the changing electrical field of the electromagnetic microwaves cause dipole rotations. These molecular dipole rotations generate friction and therefore heat between molecules. Food heats up fast and volumetrically. However, very complex interactions between microwaves and uneven food properties such as dielectric and thermal properties, produce non-uniform heating.

Coupled electromagnetic wave equations (Maxwell's equations) and heat transfer equations in space and in time can be solved iteratively to understand food molecules' chaotic dipole rotations causing friction and heating phenomena. However another difficulty arises from being able to assign appropriate dielectric and thermal properties to every food molecule that are changing continuously, but unevenly from frozen phase to liquid phase, in space and in time.

Diffusion of thermal energy from hot spots to cold spots in a food structure under the bombardment of electromagnetic microwaves can be determined by numerical solutions. The coupled heat transfer and electromagnetic wave energy equation in unsteady state for a food structure's volume element, dV=dxdydz, in a rectangular coordinate system is given in Equation 18-1.

$$\rho \times c_p \times \frac{\partial T}{\partial t} = \frac{\partial}{\partial x}\left(k\frac{\partial T}{\partial x}\right) + \frac{\partial}{\partial y}\left(k\frac{\partial T}{\partial y}\right) + \frac{\partial}{\partial z}\left(k\frac{\partial T}{\partial z}\right) + P_V(x, y, z, t) \quad \text{18-1}$$

where $\rho = \rho(x, y, z, t)$ is the density of food material at a specific location and time in kg/m^3. $c_p = c_p(x, y, z, t)$ is the specific heat at constant pressure of food material at a specific location and time in J/(kg-K). $k = k(x, y, z, t)$ is the thermal conductivity of food material at a specific location and time in W/(m-K). $P_v(x, y, z, t)$ is the electrical energy dissipated as heat in W/m^3 from microwaves entering food's volume element at a specific location and time. T is the temperature of food's volume element at a specific location and time in K. t is the food material's heating time in seconds. Equation 18-1 coupled with Maxwell's equations for microwave modeling, see Ref. [30], can provide us with reliable results backed by experiments, in understanding a microwave oven's non-uniform heating process, see Ref. [28].

A food material's heating time predictions in a microwave oven can be simplified by applying the following assumptions:

1. Only Purified water will be heated from an initial of 20 C (293 K) to 100 C (373 K).
2. Shape of the glass container that water resides in will have a spherical shape which provides uniform microwave power absorption for volumetric heating.
3. Water's dielectric and thermal properties will change during the temperature rise in the volume of water.
4. Initially, a constant value of microwave power will be dissipated by water molecules during the heating process, i.e. 1000 W. Microwave power level will not change by the size of the glass container which will vary from 0.1 liter to 2.0 liters.
5. Water with the spherical glass container will be placed at the center of oven's turntable.

With above simplifying assumptions, the coupled heat transfer energy equation in unsteady state for a volume, V, of water in m^3 is shown in Equation 18-2.

$$\rho \times V \times c_p \times \frac{\partial T}{\partial t} = P \quad \text{18-2}$$

where ρ, c_p, P are functions of water temperature during the microwave heating process and P is in Watts.

First let us examine effects of changing thermal properties during the microwave heating process of water from 20 C to 100 C. Water density versus water temperature in the region of temperature interest is shown in Figure 18-1. The variability of water density with respect to its mean density for water temperatures between 20 C and 100 C is 1.34%.

Water specific heat at constant pressure versus water temperature in the region of temperature interest is shown in Figure 18-2. The variability of water's specific heat at constant pressure with respect to its mean value between 20 C and 100 C is 0.32%. So in Equation 18-2, ρ and c_p can be assumed to be constants between 20 C and 100 C in order to have a good estimate for the time that it will take to boil water in a microwave oven.

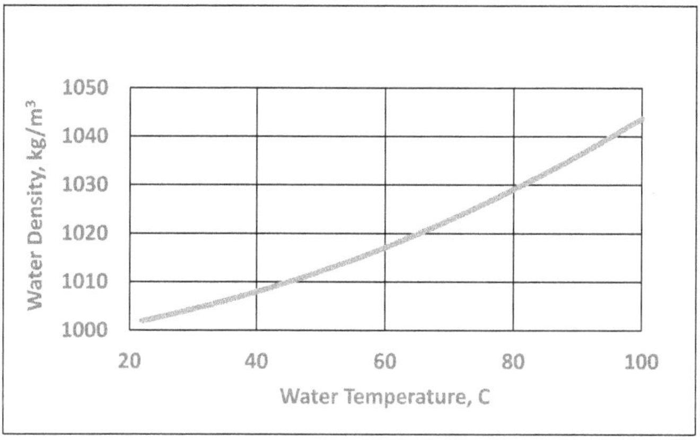

Figure 18-1 Water density versus temperature.

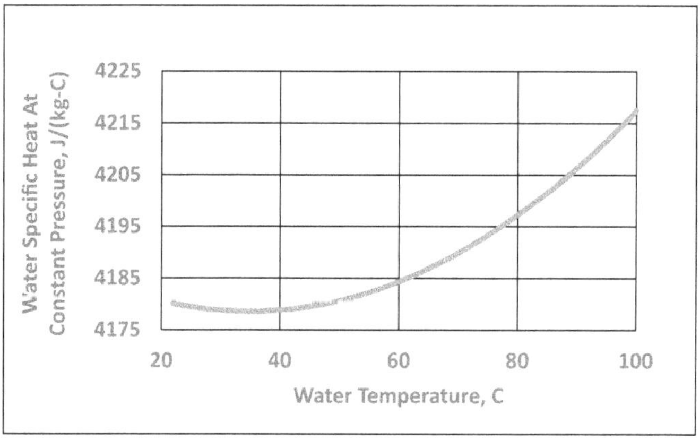

Figure 18-2 Water specific heat at constant pressure versus temperature.

Microwave power dissipated in a unit volume of water, P_v, in W/m^3 can be expressed in terms of the dielectric loss factor, ε'', of water molecules as shown Equation 18-3. See Ref. [28].

$$P_V = 2 \times \pi \times f \times \varepsilon_0 \times \varepsilon'' \times E^2 \qquad 18\text{-}3$$

where f is the microwave frequency in Hz. ε_0 is the permittivity of free space, namely 8.854×10^{-12} J/(V^2-m). The dielectric loss factor, ε'', of water's dipole molecules have a relative value which determines the ability of water to dissipate electrical energy as heat. E is the electric field strength gradient acting on water molecules in V/m.

In Equation 18-3, let us assume that all parameters are constants during water's heating process except water's dipole molecules' dielectric loss factors. Water's dielectric loss factor versus water temperature is shown in Figure 18-3. Figure 18-3 shows that as water's dipole molecules heat up, they lose their ability to dissipate more electrical energy as heat.

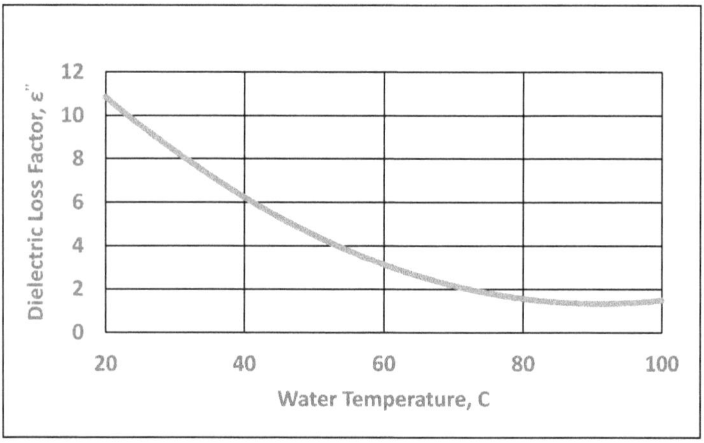

Figure 18-3 Dielectric loss factor for water's dipole molecules versus temperature.

Let us now analyze Equation 2 in three different ways. One way is to assume a constant microwave oven power of 1000 W with an average water density and an average specific heat at constant pressure all throughout the heating process. The results of water boiling time versus volume of water (dash-dot line) are shown in Figure 18-4. A second way is to assume that all dipole molecules in water are losing their ability to dissipate electrical energy as heat equally as temperature rises, namely their dielectric loss factors decrease equally as shown in Figure 18-3.

Temperature dependent decreasing microwave power case is shown in Figure 18-4 as the solid line. These water boiling times represent the worst case scenario since all water's dipole molecules are losing their ability to dissipate electrical energy as heat equally as temperature rises. This scenario provides the highest boiling times.

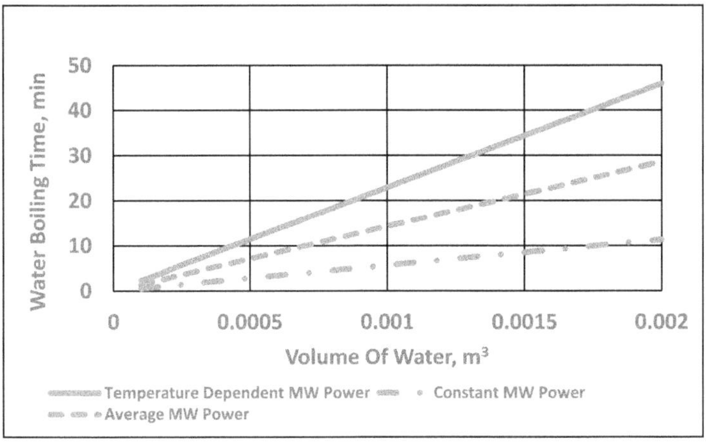

Figure 18-4 Water boiling time versus volume of water.

However, dielectric loss factor degradation for water's every dipole molecule in this water boiling event is different. Dielectric loss factor degradation is a chaotic and a probabilistic event starting at the outer surface layer of water's volume and migrating inwards as time goes on. A logical water boiling time versus volume of water curve is the average of a constant microwave power (with no dielectric loss factor degradation) and worst case scenario curves (in which all dipole molecules' dielectric loss factors degrade equally). These averaged water boiling times versus volume of water values are shown as a dash line in Figure 18.4.

19

Electromagnetic Radiation Used in Medical Diagnosis

We receive naturally occurring background electromagnetic radiation continuously. Naturally occurring radioactive materials are found in air, water and soil. We are exposed to cosmic and terrestrial radiation especially in high altitudes. Electromagnetic radiation sources are also human made such as nuclear power generation, medical uses for treatment and for diagnosis. In this chapter, I will review electromagnetic radiation exposures resulting from medical diagnosis.

There are two categories of electromagnetic radiation that are used extensively in medical diagnosis. One category is ionizing radiation which is emitted as gamma rays or x-rays during the imaging processes such as radiography (x-ray), computer tomography (CT scan uses x-rays) and nuclear medicine (uses gamma rays). The other category is non-ionizing electromagnetic radiation which is emitted as radio waves or sound waves during the imaging processes such as magnetic resonance imaging (MR) which uses radio frequency waves and ultrasound imaging which uses sound waves.

Let us first expand on medical diagnosis procedures that are using non-ionizing electromagnetic radiation.

Sound waves used in ultrasound imaging have frequencies ranging from 2 to 18 MHz, namely way above human hearing limit of 20 kHz. Sound waves used in ultrasound imaging do not harm human body cells or tissues.

Magnetic resonance imaging (MRI) uses radio frequency electromagnetic waves during the three-dimensional imaging process of the body. Standard MRI systems have up to 1.5 Tesla (1 Tesla = 10,000 Gauss) static magnetic field strength coupled with 64 MHz radio frequency waves to create images of the body. Higher magnetic fields in an MRI system produces higher resolution images of the body parts. Next level of a MRI system has 3.0 Tesla static magnetic field strength coupled with 128 MHz radio frequency waves. Some MRI systems that are used for research purposes can go up to 10 Tesla static magnetic field strength with radio frequencies as high as 400 MHz.

During the MRI imaging process, radio frequency electromagnetic waves' energy interacts with human tissues and this energy is transformed into heat. This ohmic tissue heating is maximum at the outer periphery of the human body and heating decreases towards the center of the body. Radio frequency electromagnetic waves' energy absorption by a tissue is represented by specific absorption ratio (SAR) which has the unit of W/kg. SAR is given in Equation 19-1.

$$SAR = 0.5 \times \left(\frac{\sigma}{\rho} \right) \times E^2 \qquad 19\text{-}1$$

where σ is the electrical conductivity of the tissue in S/m (Sievert per meter), ρ is the density of the tissue in kg/m^3, and E is the instantaneous electrical field acting on the tissue in V/m. Equation 19-1 can also be rewritten in terms of temperature rise in a tissue for a given exposure time as shown in Equation 19-2.

$$SAR = c_p \times \frac{\partial T}{\partial t} \qquad 19\text{-}2$$

where c_p is the specific heat at constant pressure of tissue in J/(kg-K), dT is the temperature rise in the tissue in K and ∂t is tissue's exposure time in seconds to radio frequency electromagnetic waves.

FDA recommends that for MRI imaging for an adult human body, SAR should be less than 4 W/kg in a 15 minute exposure. Using Equation 19-2 with an average specific heat of 3391 J/(kg-K) for an adult's whole body, the temperature rise during a 15 minute MRI test should not exceed 1.06 C. Similarly, FDA recommends that for MRI imaging for an adult's head, SAR should be less than 3.2 W/kg in a 10 minute exposure. Using Equation 19-2 again with an average specific heat of 3600 J/(kg-K) for an adult's head, the temperature rise during a 10 minute MRI imaging should not exceed 0.53 C. These limiting temperature rise estimates during an MRI imaging are for adult humans with normal thermo-regulatory functions. See Ref. [8].

The other category of electromagnetic radiation used in medical diagnosis is ionizing radiation which is emitted as gamma rays or x-rays during the imaging processes such as radiography (x-ray), computer tomography (CT scan uses x-rays) and nuclear medicine (uses gamma rays).

Ionizing electromagnetic waves has high energy levels and when they interact with a human tissue, they ionize atoms. Sometimes gamma rays may pass through our bodies without interacting with any tissue at all. Gamma rays have the highest energy levels, i.e. > 100,000 eV. X-rays' energy levels vary from 100 eV to 100,000 eV. When these high energy electromagnetic waves move through our body, they pass through our bones, organs and tissues differently, namely atoms disintegrate differently depending upon the type of body part that they interact with, which allows imaging of our interiors.

Risk that a person will suffer health effects from an exposure to low levels of gamma rays or x-rays radiation is measured by the Standard International (SI) unit for effective dose called Sievert (Sv). Values of Sievert characterizes a person's biological risk to effective radiation exposure dose which depends upon the ability to transfer the radiation energy to cells of the body, i.e. 1 Sv = 1 J/kg. Since radiation exposures during medical diagnosis imaging are very low, the standard unit that is used in literature is in millisievert, (mSv), which is equal to 0.001 Sv. See Ref. [10].

Gamma-rays or x-rays radiation exposure dose of 500 mSv can begin to cause symptoms of radiation poisoning. Legal annual limit to gamma-rays or x-rays radiation exposure is 50 mSv for workers in the nuclear industry, namely an order of magnitude less than the radiation poisoning limit.

People on Earth are exposed to natural sources of ionizing radiation all the time at a rate of about 3.0 mSv/year. These natural resources of ionizing radiation come from naturally occurring radioactive materials in the ground, soil and water, from cosmic radiation, from radon, etc. Exposure to cosmic rays during a round trip flight between Los Angeles and New York can be about 0.03 mSv.

Typical effective ionizing radiation doses received from diagnostic x-rays is given in Table 19-1. Typical effective ionizing radiation doses received from imaging during interventional procedures are listed in Table 19-2. See Ref. [10].

Table 19-1 Typical effective ionizing radiation doses received from diagnostic x-ray imaging.

X-ray Imaging Exam	Effective Dose, mSv
Chest	0.1
Lumbar Spine	1.5
Abdomen or Hip	0.6
Mammogram (2 views)	0.36
Dental (panoramic)	0.01
Skull	0.1
Hand or Foot	0.005

Nuclear medicine uses very small amounts of radioactive materials called radiotracers which are taken internally. Then, external gamma ray cameras capture and form images from the radiation emitted by the radiotracers. A nuclear medicine examination's effective ionizing radiation dose can be high, i.e. as high as 20 mSv.

Table 19-2 Typical effective ionizing radiation doses received from imaging during interventional procedures.

X-ray Imaging During Interventional Procedure	Effective Dose, mSv
Upper Gastro-Intestinal	6.0
Abdomen Kidney, Ureter, Bladder	0.7
CT Head	2.0
CT Chest	7.0
CT Abdomen or Pelvis	10.0
Whole Body CT Screening	10.0
CT Biopsy	1.0
Coronary Angiography	20.0
Cardiac Diagnostic and Intervention	30.0
Noncardiac Embolization	55.0

20

Storing Liquid Oxygen in a Cryogenic Spherical Container

Liquid oxygen is used widely in industrial and medical fields and also in spacecraft rocket propellants. Liquid oxygen storage is less bulky and less costly than the equivalent capacity of high pressure gas oxygen storage. The freezing point of liquid oxygen in one atmosphere of pressure is –219 C (54 K). The boiling point of liquid oxygen in one atmosphere of pressure is –183 C (90 K). In this chapter, let us investigate the radiation heat transfer from a cryogenic spherical container for storing liquid oxygen. The spherical cryogenic container has a concentric outer spherical shell and facing spherical surfaces "1" and "2" are silver coated in order to minimize radiation heat losses from the liquid oxygen as shown in Figure 20-1.

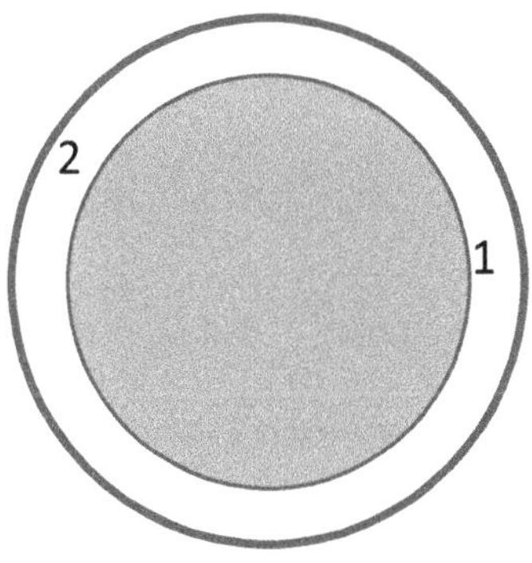

Figure 20-1 Liquid oxygen cryogenic spherical container whose outside surface is designated as surface "1" with a concentric outer spherical shell whose inside surface is designated as "2". There is vacuum in the space between the two spheres.

Let us assume that surface "2" is kept at 0 C (273 K) all the time. Initially, the liquid oxygen in the cryogenic sphere is at - 200 C (73 K) which is assumed to be the same temperature as surface "1". We will calculate the time it will take for the liquid oxygen's temperature, i.e. surface "1" temperature, to reach -183 C (90 K) for different emissivity values for surfaces "1" and "2". During this radiative heating period, it is assumed that liquid oxygen has a constant density of 1141 kg/m^3 and a constant specific heat at constant pressure of 921 J/(kg-K). In the present analysis, conduction heat transfer losses are assumed to be negligible in the walls of the concentric spheres, namely small wall thicknesses are assumed as compared to diameters of the concentric spheres.

Radiation heat transfer under steady-state conditions from surface "1" to surface "2" is given in Watts in Equation 20-1. In the present analysis, heat losses from fill and gauge lines are neglected.

$$Q_{1\to 2} = \frac{A_1 \times \sigma \times \left(T_1^4 - T_2^4\right)}{\sum R_{1\to 2}} \qquad \text{20-1}$$

where A_1 is the surface area of "1" in m^2. σ is the Stefan-Boltzmann constant at 5.67×10^{-8} $W/(m^2\text{-}K^4)$. T_1 is the liquid oxygen temperature or surface "1" temperature in K. T_2 is the surface "2" temperature at 273 K. Sum of resistances for radiation heat transfer between the two diffuse and gray surfaces "1" and "2" and which have vacuum between them, are given in Equation 20-2. See Ref. [33].

$$\sum R_{1\to 2} = \left(\frac{1-\varepsilon_1}{\varepsilon_1}\right) + \left(\frac{1}{F_{1\to 2}}\right) + \left(\frac{1-\varepsilon_2}{\varepsilon_2}\right) \times \left(\frac{A_1}{A_2}\right) \qquad \text{20-2}$$

where ε_1 and ε_2 are emissivities of surfaces "1" and "2", respectively. $F_{1\to 2}$ is the configuration factor from surface "1" to surface "2" and its value is 1.0. A_2 is the surface area "2" in m^2. Equation 20-2 can be simplified and rewritten as shown in Equation 20-3.

$$\sum R_{1\to 2} = \left(\frac{1}{\varepsilon_1}\right) + \left(\frac{1-\varepsilon_2}{\varepsilon_2}\right) \times \left(\frac{D_1^2}{D_2^2}\right) \qquad \text{20-3}$$

where D_1 is the diameter of surface "1" and D_2 is the diameter of surface "2" in meters.

Times for liquid nitrogen to heat from 73 K to 90 K versus surface emissivities of spheres' materials are shown in Figure 20-2. $\varepsilon_1 = \varepsilon_2$ in the present calculations. Also $D_1 = 0.5$ *m* and $D_2 = 0.6$ *m*. Liquid oxygen heating times decrease fast for increasing emissivity values at low surface emissivities. Decreasing heating times slow down appreciably above surface emissivities of 0.05. Sensitivity of change in time for liquid oxygen to heat from 73 K to 90 K to change in surface emissivities is shown in Figure 20-3. Absolute values of this sensitivity at low surface emissivities are very high. Absolute values of this sensitivity approaches zero fast above the surface emissivity value of 0.06.

Using steady-state heat transfer calculations versus transient heat transfer numerical calculations only makes 0.67 % difference in the heat lost from the liquid oxygen at t=0 when T1=73 K and at t=65.14 hours when T1=90 K for $\varepsilon_1=\varepsilon_2=0.03$. So steady-state heat transfer calculations give fairly accurate results in the present analysis.

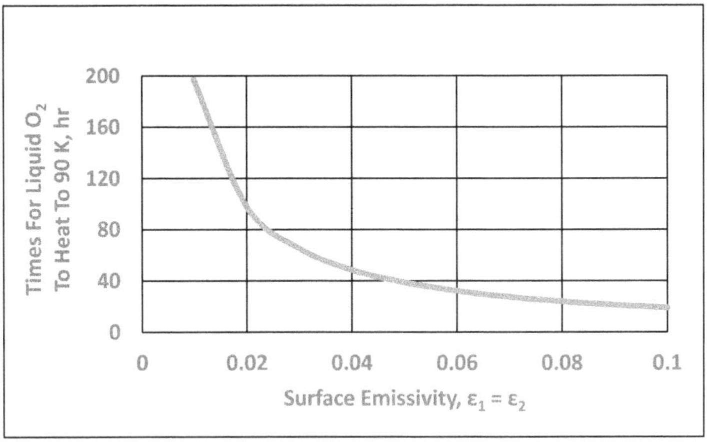

Figure 20-2 Times for liquid nitrogen to heat from 73 K to 90 K versus surface emissivities. $\varepsilon_1=\varepsilon_2$ in calculations.

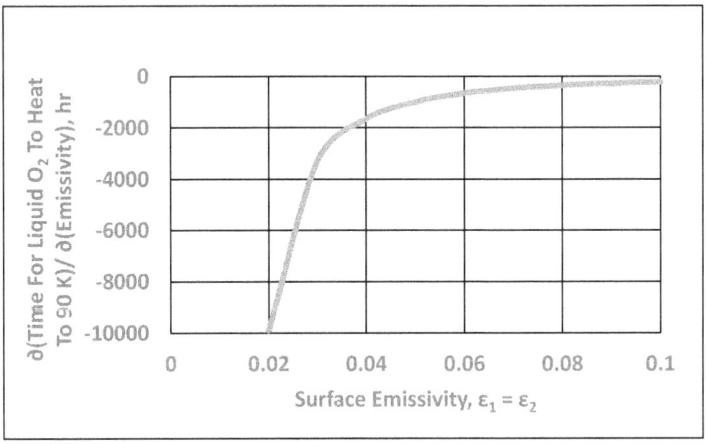

Figure 20-3 Sensitivity of change in time for liquid oxygen to heat from 73 K to 90 K to change in surface emissivities. $\varepsilon_1=\varepsilon_2$ in calculations.

Time for liquid oxygen to heat to from 73 K to 90 K versus increasing inner diameter of "2" is shown in Figure 20-4. $\varepsilon_1 = \varepsilon_2 = 0.03$ in these calculations. The diameter "1" is fixed at 0.5 meter in the following calculations. As diameter "2" increases, total radiation heat transfer resistance decreases and therefore the radiation heat loss from the liquid oxygen increases. So times for liquid oxygen to heat from 73 K to 90 K decreases with increasing diameter "2".

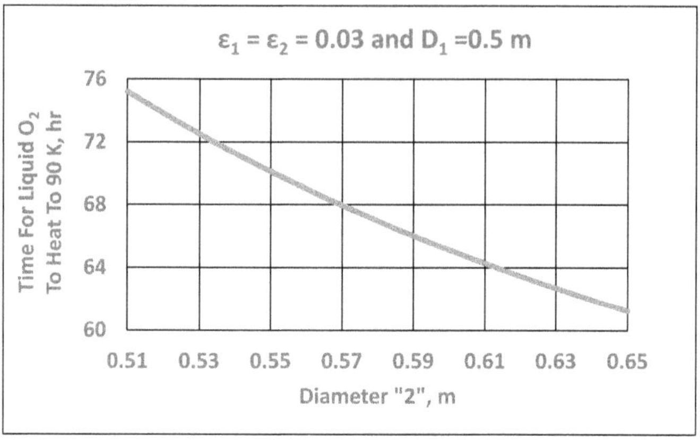

Figure 20-4 Time for liquid oxygen to heat to from 73 K to 90 K versus diameter "2". $\varepsilon_1=\varepsilon_2 = 0.03$ in calculations.

Sensitivity of change in time for liquid oxygen to heat from 73 K to 90 K to change in diameter "2" is shown in Figure 20-5. $\varepsilon_1 = \varepsilon_2 = 0.03$ in these calculations. Absolute values of this sensitivity are high when diameter "2" is close to 0.5 meter. Absolute values of this sensitivity decrease as the diameter of "2" increases.

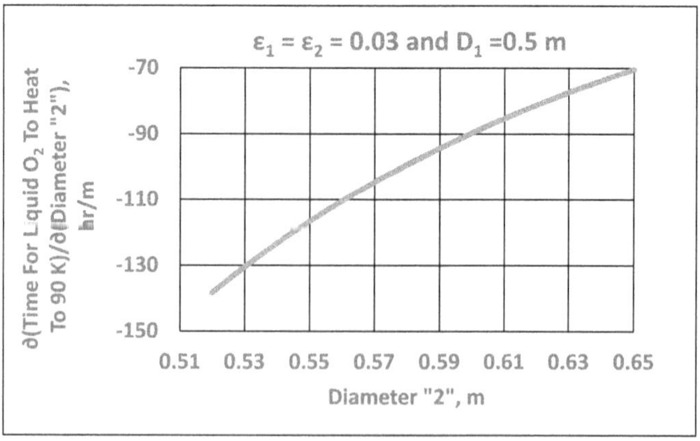

Figure 20-5 Sensitivity of change in time for liquid oxygen to heat from 73 K to 90 K to change in diameter "2". $\varepsilon_1=\varepsilon_2 = 0.03$ in calculations.

Radiation heat transfer from surface "1", to surface "2" versus diameter "2" is shown in Figure 20-6. ε_1=ε_2= 0.03 in these calculations. Absolute value of radiation heat transfer from surface "1" increases as diameter "2" gets larger and radiation heat transfer resistance from surface "1" to surface "2" decreases.

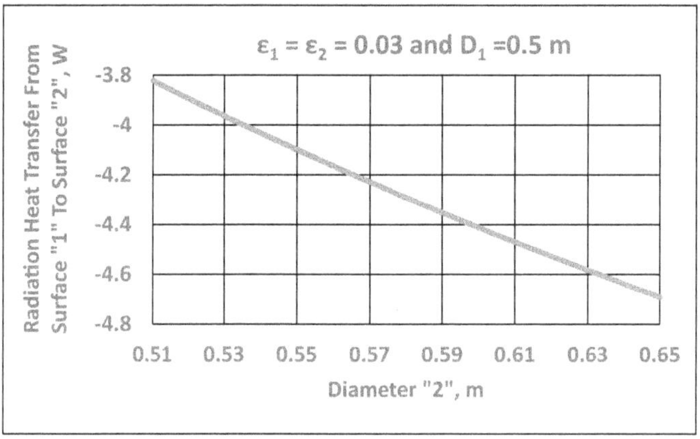

Figure 20-6 Radiation heat transfer from surface "1" to surface "2" versus diameter "2". ε_1=ε_2 = 0.03 in calculations.

Total resistance, $\left(\frac{1}{A_1}\right) \times \sum R_{1\to 2}$, from the surface "1" to surface "2" versus diameter "2" is shown in Figure 20-7. ε_1=ε_2 = 0.03 in these calculations. A 10% increase in diameter of "2" from 0.51 m to 0.561 m decreases the total radiation heat transfer resistance by 8.3% and therefore increases the radiation heat transferred out from surface "1" by 8.3%.

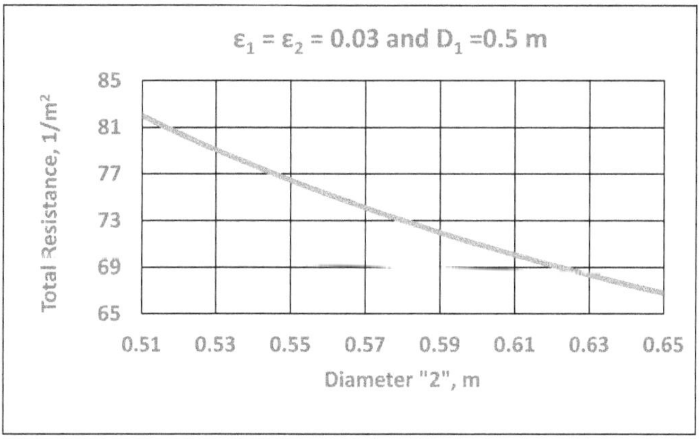

Figure 20-7 Total radiation heat transfer resistance from surface "1" to surface "2" versus diameter "2". ε_1=ε_2 = 0.03 in calculations.

21

Storing Liquid Oxygen in a Cryogenic Spherical Container with a Radiation Shield

Storing liquid oxygen in a cryogenic spherical container with a concentric spherical outer shell was analyzed in Chapter 20. In this chapter, let us insert a concentric spherical radiation shield with low emissivity surfaces in between the cryogenic spherical container "1" and outer spherical shell "2" in order to minimize radiation heat losses from the liquid oxygen as shown in Figure 21-1.

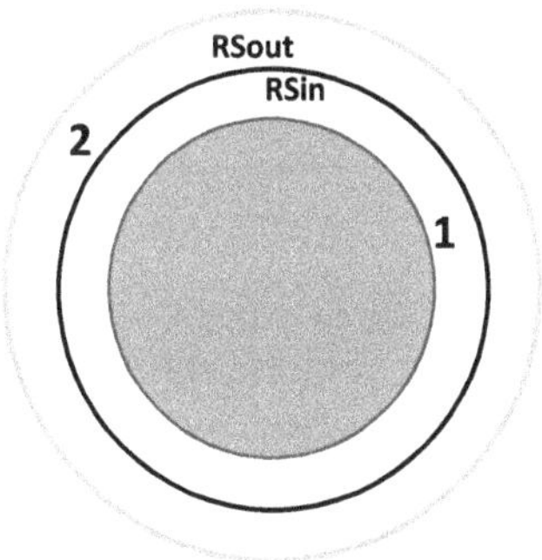

Figure 21-1 Liquid oxygen cryogenic spherical container whose outside surface is designated as surface "1" with a concentric outer spherical shell whose inside surface is designated as "2". There is vacuum in the space between the two spheres. There is also a spherical radiation shield between "1" and "2" where "RSin" is the inner spherical surface of the radiation shield facing surface "1" and "RSout" is the outer spherical surface of the radiation shield facing surface "2".

In the present analysis, surface "2" is kept at 0 C (273 K) all the time. Initially, the liquid oxygen in the cryogenic sphere is at –200 C (73 K) which is assumed to be the same temperature as surface "1". We will calculate the time it will take for the liquid oxygen's temperature, i.e. surface "1" temperature, to reach –183 C (90 K) for different emissivity values for the inner and outer surfaces of the radiation shield. Also, it is assumed that inner and outer surfaces of the radiation shield have equal emissivities. In the present analysis, it is assumed that $\varepsilon_1=\varepsilon_2=0.05$. During this radiative heating period, it is assumed that liquid oxygen has a constant density of 1141 kg/m^3 and a constant specific heat at constant pressure of 921 J/(kg-K). In the present analysis, conduction heat transfer losses are assumed to be negligible in the walls of the concentric spheres and the radiation shield, namely small wall thicknesses are assumed as compared to diameters of the concentric spheres and the radiation shield.

Radiation heat transfer under steady-state conditions from surface "1" to surface "2" is given in Watts in Equation 21-1.

$$Q_{1\to 2} = \frac{A_1 \times \sigma \times \left(T_1^4 - T_2^4\right)}{\sum R_{1\to 2}} \qquad 21\text{-}1$$

where A_1 is the surface area of "1" in m^2. σ is the Stefan-Boltzmann constant at 5.67x10^{-8} W/(m^2-K^4). T_1 is the liquid oxygen temperature and the surface "1" temperature in K. T_2 is the surface "2" temperature at 273 K. Sum of resistances for radiation heat transfer between the four diffuse and gray surfaces "1", "RS$_{in}$", "RS$_{out}$", and "2" and which have vacuum between them, are given in Equation 21-2. See Ref. [33].

$$\left(\frac{1}{A_1}\right)\sum R_{1\to 2} = \left(\frac{1-\varepsilon_1}{\varepsilon_1}\right)\times\left(\frac{1}{A_1}\right) + \left(\frac{1}{A_1 \times F_{1\to RSin}}\right) + \left(\frac{1-\varepsilon_{RSin}}{\varepsilon_{RSin}}\right)\times\left(\frac{1}{A_{RSin}}\right) +$$
$$\left(\frac{1-\varepsilon_{RSout}}{\varepsilon_{RSout}}\right)\times\left(\frac{1}{A_{RSout}}\right) + \left(\frac{1}{A_{RSout}\times F_{RSout\to 2}}\right) + \left(\frac{1-\varepsilon_2}{\varepsilon_2}\right)\times\left(\frac{1}{A_2}\right) \qquad 21\text{-}2$$

where ε_1, ε_2, ε_{RSin} and ε_{RSout} are emissivities of surfaces "1", "2" and radiation shield's inner and outer surfaces, respectively. A_2 is the surface area "2" in m^2. A_{RSin} and A_{RSout} are the inner and the outer surface areas of the radiation shield in m^2, respectively. Also, configuration factors $F_{1\to RSin}$ and $F_{RSout\to 2}$ are equal to one and $\varepsilon_{RSin} = \varepsilon_{RSout} = \varepsilon_{RS}$. Then sum of resistances for radiation heat transfer between the four diffuse and gray surfaces in Equation 21-2 simplifies to Equation 21-3 as shown below.

$$\sum R_{1\to 2} = \frac{1}{\varepsilon_1} + \left(\frac{1-\varepsilon_{RS}}{\varepsilon_{RS}}\right)\times\left(\frac{A_1}{A_{RSin}} + \frac{A_1}{A_{RSout}}\right) + \left(\frac{A_1}{A_{RSout}}\right) + \left(\frac{1-\varepsilon_2}{\varepsilon_2}\right)\times\left(\frac{A_1}{A_2}\right) \qquad 21\text{-}3$$

Radiation heat transfer from liquid oxygen to outer spherical shell versus radiation shield's inner and outer surfaces' emissivities where $\varepsilon_{RSin}=\varepsilon_{RSout}$, is shown in Figure 21-2. In these calculations, spherical diameters D_1=0.5 m, D_2=0.6 m, D_{RSin}=0.549 m and D_{RSout}=0.551 m. Also $\varepsilon_1=\varepsilon_2$=0.05 in the present calculations. Radiation heat transferring out of liquid oxygen increases fast as the radiation shield's surface emissivity increases. Value of radiation heat transfer approaches 7.41 Watts as the radiation shield's surface emissivity approaches 1.0 which is the radiation heat lost from liquid oxygen cryogenic container without the radiation shield.

Times for liquid oxygen to heat to from 73 K to 90 K versus radiation shield's inner and outer surfaces' emissivities where $\varepsilon_{RSin}=\varepsilon_{RSout}$ are shown in Figure 21-3. Time for liquid oxygen to heat to from 73 K to 90 K approaches 38.8 hours which is the heating time for the liquid oxygen without the radiation shield as emissivities of radiation shield's surfaces approach 1.0.

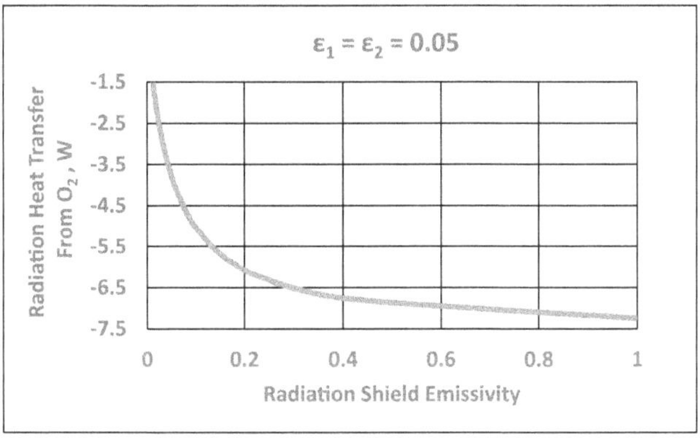

Figure 21-2 Radiation heat transfer from liquid oxygen to outer spherical shell versus radiation shield's inner and outer surfaces' emissivities where $\varepsilon_{RSin}=\varepsilon_{RSout}$.

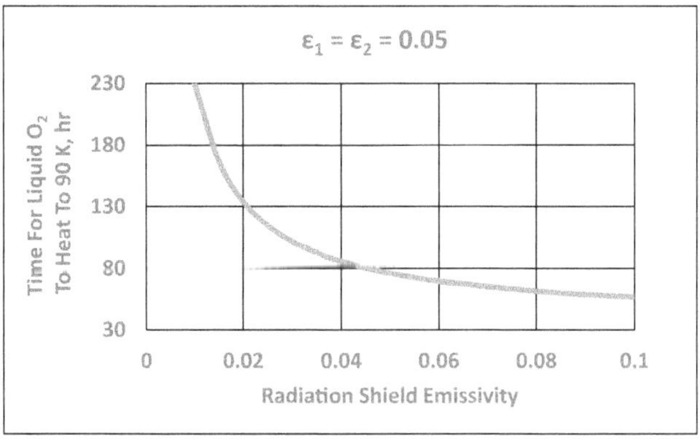

Figure 21-3 Time for liquid oxygen to heat from 73 K to 90 K versus radiation shield's inner and outer surfaces' emissivities where $\varepsilon_{RSin}=\varepsilon_{RSout}$.

Change in time for liquid oxygen to heat from 73 K to 90 K to change in radiation shield's inner and outer surfaces' emissivities where $\varepsilon_{RSin}=\varepsilon_{RSout}$, is shown in Figure 21-4. Absolute value of sensitivity of change in time for liquid oxygen to heat from 73 K to 90 K to changing radiation shield's surface emissivity is very high at low radiation shield emissivities, i.e. around $\varepsilon_{RSin}=\varepsilon_{RSout}=0.02$. Absolute value of this sensitivity goes to zero as $\varepsilon_{RSin}=\varepsilon_{RSout} > 0.1$.

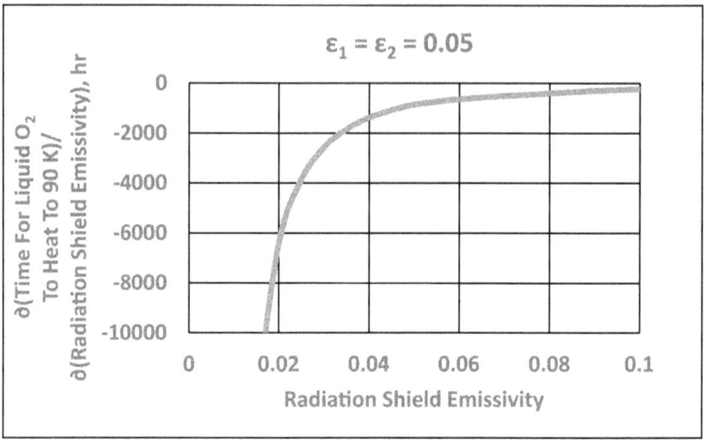

Figure 21-4 Sensitivity of change in time for liquid oxygen to heat from 73 K to 90 K to change in radiation shield's inner and outer surfaces' emissivities where $\varepsilon_{RSin}=\varepsilon_{RSout}$.

22

Temperatures of Different Surfaces Exposed to Incident Extraterrestrial Radiation at Several Locations

In this chapter, temperatures of different horizontal surfaces, i.e. angle between the plane of the surface and the horizontal is zero degree, are determined at 21st of June, i.e. 172nd day of the year, at 12:00 noon solar time of the day, i.e. solar hour angle is zero degree, namely angular displacement of the Sun east or west of the local meridian due to rotation of Earth on its axis is zero degree, and at four different locations. The locations are Anchorage, Alaska at 61.17 degrees latitude north, Seattle, Washington at 47.61 degrees latitude north, Sacramento, California at 38.58 degrees latitude north and San Diego, California at 32.73 degrees latitude north. All four locations are very close to the sea level, namely altitudes vary between 9 meters and 35 meters. So altitude effects are neglected in the following calculations.

Three type of horizontal surfaces whose back sides are well insulated are analyzed. These surfaces have white paint or black paint on them or made out of red brick. They are diffuse and gray surfaces and their absorptivities, α, and emissivities, ε, are given below:

White paint on metallic surface:	$\alpha = 0.24$,	$\varepsilon = 0.96$
Black paint on metallic surface:	$\alpha = 0.97$,	$\varepsilon = 0.97$
Red brick surface:	$\alpha = 0.63$,	$\varepsilon = 0.93$

Extraterrestrial radiation reaches a surface by direct beam radiation, by diffuse radiation and by ground reflected radiation. Heat lost from the surface are assumed to be by infrared radiation and by convection on a calm day. Net heat transfer, Q_{net}, in W/m^2 at the horizontal surface under steady-state conditions is given in Equation 22-1.

$$Q_{net} = \alpha \times Q_{dbr} + \varepsilon \times Q_{dr} + \varepsilon \times Q_{grr} - Q_{er} - Q_{conv} \qquad \text{22-1}$$

where Q_{dbr} is the direct beam extraterrestrial radiation in W/m^2 that irradiates on to the horizontal surface under clear sky conditions at sea level and it is defined in Equation 22-2.

$$Q_{dbr} = Q_{sc} \times \left[1 + 0.033 \times cos\left(\frac{360 \times N}{365} \right) \right] \times AT \times cos(\theta_z) \qquad \text{22-2}$$

where Q_{sc} is the solar constant (extraterrestrial radiation received mainly from the Sun) with a value of 1,360.8 W/m^2 at the outside of the atmosphere and at the mean Earth to Sun distance. N is the day of the year and it is 172 in the present calculations. AT is the atmospheric transmittance that covers effects of scattering and absorption of extraterrestrial radiation in the atmosphere. θ_Z is the zenith angle between the horizontal plane and the line to the Sun.

The atmospheric transmittance can estimated at sea level by the following Equations 22-3a and 22-3b. See Ref. [6].

$$AT = 0.124 + 0.749 \times exp\left[\frac{-0.395}{cos(\theta_z)} \right] \qquad \text{22-3a}$$

$$AT = 0.127 + 0.749 \times exp\left[\frac{-0.391}{cos(\theta_z)} \right] \qquad \text{22-3b}$$

Equation 22-3a is used for mid-latitude summer, namely for Seattle, Sacramento and San Diego in the present calculations. Equation 22-3b is used for sub-arctic summer, namely for Anchorage in the present calculations.

The zenith angle for horizontal surfaces can be determined by using the following Equation 22-4. See Ref. [6].

$$cos(\theta_z) = cos(\varphi) \times cos(\delta) \times cos(\omega) + sin(\varphi) \times sin(\delta) \qquad \text{22-4}$$

where φ is the latitude angle for the location in question and north is positive. δ is declination angle which defined in Equation 22-5 and it is the angle between light rays from the Sun and Earth's equator. ω is the hour angle and it is zero in the present calculations since calculations are done for 12:00 noon solar time of the day. See Ref. [6].

$$\delta = 23.45 \times sin\left[360 \times \left(\frac{284 + N}{365}\right)\right] \qquad 22\text{-}5$$

Q_{dr} is the diffuse component of hourly extraterrestrial radiation on a clear day in W/m^2 and it can be estimated by the fraction of the hourly extraterrestrial radiation on a horizontal surface given in Equation 22-6. See Ref. [6].

$$Q_{dr} = 0.165 \times Q_{dbr} \qquad 22\text{-}6$$

Q_{dr} diffuse radiation energy from the sky is assumed to be in the infrared region as the surface emission radiation energy from the horizontal surface and therefore absorptivity of Q_{dr} is assumed to be the same as the surface emissivity ε.

Q_{grr} is the ground reflected component of the hourly extraterrestrial radiation on a clear day in W/m^2 and it is zero for the present calculations for a horizontal surface. Q_{grr} radiation energy is assumed to be in the infrared region as the surface emission radiation energy and therefore absorptivity of Q_{grr} is assumed to be the same as the surface emissivity ε.

Q_{er} is the radiation emitted from the diffuse and gray horizontal surface in the infrared region of the electromagnetic spectrum in W/m^2 and it is given in Equation 22-7.

$$Q_{er} = \varepsilon \times \sigma \times \left(T_s^4 - T_{env}^4\right) \qquad 22\text{-}7$$

where σ = 567×10^{-8} $W/(m^2\text{-}K^4)$, T_s is the temperature of the horizontal surface in K and T_{env} is the environment's temperature which is at 293 K in the present calculations.

Q_{conv} is the convection heat transfer from the top surface of the horizontal surface to the environment in W/m^2. The heat transfer coefficient, *h*, for a calm day can be estimated by Equation 22-8. See Ref. [13].

$$h = 0.22 \times \left(T_s - T_{env}\right)^{1/3} \qquad 22\text{-}8$$

Then the convection heat transfer from the top surface of the horizontal surface to the environment is shown in Equation 22-9.

$$Q_{conv} = 0.22 \times \left(T_s - T_{env}\right)^{4/3} \qquad 22\text{-}9$$

Equations 22-1 through 22-9 are used under equilibrium conditions, namely when Q_{net} = 0, to calculate the temperature of the horizontal surface for three different surface finishes and for four different locations. Since Equation 22-1 provides a non-linear equation in T_s when Q_{net} = 0, every case has to be solved by trial and error for the surface temperature.

First let us calculate Q_{dbr} which is the direct beam extraterrestrial radiation on a horizontal surface at a given location at a certain day of the year, namely N = 172 days, and at a certain time of the day, namely 12:00 noon solar time or the hour angle ω = 0. Calculations for Q_{dbr} are shown in Table 22-1.

Table 22-1 Direct beam extraterrestrial radiation calculations on a horizontalsurface for four different locations.

Location	N, Day Of The Year	Latitude North, ϕ	Declination Angle δ, Eqn. 22-5	Hour Angle ω, 15° Per Hour, Afternoon Positive
Anchorage, AK	172	61.17	23.4497	0
Seattle, WA	172	46.61	23.4497	0
Sacramento, CA	172	38.58	23.4497	0
San Diego, CA	172	32.73	23.4497	0

Location	COS(θz) Eqn. 22-4	Zenith Angle, θz	Solar Constant, Qsc, W/m²	Atmospheric Transmittance, AT, Eqns. 22-3a & 22-3b
Anchorage, AK	0.791	37.72	1360.8	0.584
Seattle, WA	0.912	24.16	1360.8	0.610
Sacramento, CA	0.965	15.13	1360.8	0.622
San Diego, CA	0.987	9.28	1360.8	0.626

Location	Direct Beam Extraterrestrial Radiation, Q_{dbr}, W/m², Eqn. 22-2
Anchorage, AK	608.1
Seattle, WA	733.2
Sacramento, CA	790.6
San Diego, CA	814.0

Q_{dr}, the diffuse component of hourly extraterrestrial radiation on a clear day and on a horizontal surface is calculated using Equation 22-6. Q_{grr} is the ground reflected component of the hourly extraterrestrial radiation on a clear day and it is zero for a horizontal surface. Q_{dr} and Q_{grr} are shown in Table 22-2.

Table 22-2 The diffuse component of hourly extraterrestrial radiation on a clear day and on a horizontal surface and the ground reflected component of the hourly extraterrestrial radiation on a clear day and on a horizontal surface at four different locations.

Location	Q_{dr}, W/m² Eqn. 22-6	Q_{grr}, W/m²
Anchorage, AK	100.3	0
Seattle, WA	121.0	0
Sacramento, CA	130.4	0
San Diego, CA	134.3	0

Equations 22-1, 22-7 and 22-9 can be used under equilibrium conditions, namely when $Q_{net} = 0$, to calculate by iteration the temperature of a horizontal surface for three different surface finishes and for four different locations as shown in Table 22-3.

Table 22-3 Temperature of a horizontal surface for three different surface finishes and for four different locations under equilibrium conditions.

Location	Temperature Of Horizontal Surface With White Paint On Metallic Substrate, °C	Temperature Of Red Brick Horizontal Surface, °C	Temperature Of Horizontal Surface With Black Paint On Metallic Substrate, °C
Anchorage, AK	41.4	69.9	87.8
Seattle, WA	47.8	79.8	99.6
Sacramento, CA	50.6	84.0	104.6
San Diego, CA	51.7	85.7	106.6

Equilibrium temperatures on a horizontal surface increase as the location moves from the arctic towards the equator. Least equilibrium temperature increase from Anchorage to San Diego is when the surface is finished with white paint on a metallic substrate, namely 10.3 C. Most equilibrium temperature increase from Anchorage to San Diego is when the surface is finished with black paint on a metallic substrate, namely 18.8 C.

Table 22-4 shows total extraterrestrial radiation absorbed by different surface finishes at different locations. A horizontal surface with black paint on metallic substrate has the highest extraterrestrial radiation absorbed as expected, a brick surface is in the mid-range and a white paint on metallic surface has the least extraterrestrial radiation absorbed.

Table 22-4 Total extraterrestrial radiation absorbed by different surface finishes at different locations.

Location	Horizontal Surface With White Paint On Metallic Substrate, Total Extraterrestrial Radiation Absorbed, W/m²	Horizontal Brick Surface, Total Extraterrestrial Radiation Absorbed, W/m²	Horizontal Surface With Black Paint On Metallic Substrate, Total Extraterrestrial Radiation Absorbed, W/m²
Anchorage, AK	242.3	476.4	687.2
Seattle, WA	292.1	574.4	828.5
Sacramento, CA	315.0	619.4	893.4
San Diego, CA	324.3	637.8	919.9

Tables 22-5 through 22-7 show radiation emitted in the infrared region from a horizontal surface, natural convection heat lost in calm air and the natural convection heat transfer coefficients for each surface type and at each location under equilibrium conditions. 5% to 10% of the heat lost from a horizontal surface is from convection heat transfer on a calm day. 90% to 95% of the heat lost from a horizontal surface is from radiation emitted in the infrared region to the environment.

Table 22-5 Radiation emitted in the infrared region, natural convection heat lost in calm air and the natural convection heat transfer coefficients for a horizontal surface with white paint on metallic substrate.

Location	Radiation Emitted, W/m², Eqn. 22-7	Convection Heat Loss, W/m², Eqn. 22-9	Natural Convection Heat Transfer Coefficient, W/(m²-K), Eqn. 22-8
Anchorage, AK	229.2	13.1	0.611
Seattle, WA	273.6	18.5	0.666
Sacramento, CA	293.9	21.1	0.688
San Diego, CA	302.2	22.1	0.696

Table 22-6 Radiation emitted in the infrared region, natural convection heat lost in calm air and the natural convection heat transfer coefficients for a horizontal brick surface.

Location	Radiation Emitted, W/m², Eqn. 22-7	Convection Heat Loss, W/m², Eqn. 22-9	Natural Convection Heat Transfer Coefficient, W/(m²-K), Eqn. 22-8
Anchorage, AK	436.0	40.4	0.810
Seattle, WA	523.0	51.4	0.860
Sacramento, CA	563.1	56.3	0.880
San Diego, CA	579.5	58.3	0.888

Table 22-7 Radiation emitted in the infrared region, natural convection heat lost in calm air and the natural convection heat transfer coefficients for a horizontal surface with black paint on metallic substrate.

Location	Radiation Emitted, W/m², Eqn. 22-7	Convection Heat Loss, W/m², Eqn. 22-9	Natural Convection Heat Transfer Coefficient, W/(m²-K), Eqn. 22-8
Anchorage, AK	626.4	60.8	0.897
Seattle, WA	753.2	75.3	0.946
Sacramento, CA	811.7	81.7	0.966
San Diego, CA	835.6	84.3	0.973

23

Cell Phone's Radio Frequency Electromagnetic Radiation Effects on Human Brain Tissue

Cell phones and digital cellular networks can produce high enough non-ionizing radio frequency electromagnetic radiation energy during long exposures, i.e. more than 1 hour, that can increase temperature of brain tissues high enough to affect the human nervous system. International guidelines allow a localized maximum exposure which is specified as 10 W/kg specific absorption rate (SAR), averaged over a 10-gram tissue for non-ionizing radio frequency electromagnetic radiation energy. See Ref. [18]. Radio frequency radiation includes radio waves and microwaves. In the new fifth generation (5G) wireless technology for cell phones and for digital cellular networks, electromagnetic radiation waves operate between 30 and 300 GHz frequency which can produce high radiation energy levels and damage human tissues.

One way to avoid damage from usage of a cell phone is to keep it away from your body about 30 cm or more during usage because radio frequency radiation exposure falls off rapidly with distance. Use the cell phone for texting and for browsing the internet. Also use a hands free headset and avoid long and numerous voice calls with the cell phone placed against the ear. See Ref. [32].

Conservation of energy at any instant in time to a brain tissue that is just inside the human skull located behind the ear can be expressed by Equation 23-1. Equation 23-1 is a one-dimensional transient heat transfer model for the brain tissue that is just inside the human skull located behind the ear. Equation 23-1 neglects the heat gained due to metabolism.

For a more detailed three-dimensional model that couples heat transfer equations with Maxwell's equations for electromagnetic waves, refer to Ref. [4].

$$\rho_t \times c_{pt} \times \frac{dT_t}{d\theta} = -h_T \times \left(\frac{1}{L}\right) \times \left(T_t - T_{env}\right) - \varepsilon \times \sigma \times \left(\frac{1}{L}\right) \times \left(T_t^4 - T_{env}^4\right) - \left(\frac{V_b}{V_t}\right) \times \rho_b \times c_{pb} \times \omega_b \times \left(T_t - T_b\right) + Q_{EM} \quad \text{23-1}$$

where subscript "t" stands for tissue, "env" is for environment and "b" is for blood. ρ is density in kg/m^3. c_p is specific heat at constant pressure in J/(kg-K). T is temperature in K. θ is time in seconds. h_T is the total heat transfer coefficient (convection plus conduction) from outside air to brain tissues next to the skull bone in W/(m^2-K). L is the characteristic length in meters, i.e. heat transfer volume divided by the heat transfer surface area in the region considered for Equation 23-1. ε is the skin emissivity in the infrared region for a diffuse and gray surface. $\sigma = 5.67 \times 10^{-8}$ W/(m^2-K^4) is the Stefan-Boltzmann constant. $\left(\frac{V_b}{V_t}\right) = 0.57$ is the average volume of blood in the brain tissue divided by the average volume of brain tissue obtained from 55 mL of blood per 100 gm of tissue. $\omega = 0.0004$ is the blood perfusion rate in 1/s and it is assumed to be a constant in the present model. Q_{EM} is the power of radio frequency electromagnetic radiation in W/m^3.

Equation 23-1 is solved numerically using Equation 23-2 along with the initial condition at $\theta = 0$, $T_t = 310.15\ K$ *or* $37.0\ C$.

$$T_{t-(j)} = T_{t-(j-1)} + \left(\frac{\theta_{(j)} - \theta_{(j-1)}}{\rho_t \times c_{pt}}\right) \times \sum_i Q_{i-(j-1)} \quad \text{23-2}$$

Parameters and supplementary equations that are used to solve Equation 23-1 are listed below. h_T, the total heat transfer coefficient from the outside air to the brain tissue right inside the skull's bone and it is given in Equation 23-3.

$$\rho_t = 1030\ kg/m^3,\ c_{pt} = 3680\ J/(kg - K),$$

$$h_T = \frac{1}{\frac{1}{h_{conv}} + \frac{L_{skin}}{k_{skin}} + \frac{L_{fat}}{k_{fat}} + \frac{L_{bone}}{k_{bone}}} \quad \text{23-3}$$

where $h_{conv} = 1\ W/(m^2 - K)$ is the convection heat transfer coefficient between calm outside air and the skin. Thermal conductivities and conduction lengths for skin, fat and bone layers in head's skull are given below.

$k_{skin} = 0.266\ W/(m - K),\ L_{skin} = 0.001\ m,$

$k_{fat} = 0.223\ W/(m - K),\ L_{fat} = 0.001\ m,$

$k_{bone} = 0.4345\ W/(m - K),\ L_{bone} = 0.005\ m,$

$L = 0.007\ m,\ \varepsilon = 0.95,\ T_{env}\ varies\ from\ 273.15\ to\ 323.15\ K\ (0\ to\ 50\ C)$

$\rho_b = 1060\ kg/m^3,\ c_{pb} = 3840\ J/(kg - K),$

Q_{EM} is the power of radio frequency electromagnetic radiation coming from the cell phone and it is obtained from Equation 23-4.

$$Q_{EM} = 0.5 \times \rho_s \times \left(\frac{\sigma_t}{\sigma_s} \right) \times\ SAR \qquad 23\text{-}4$$

where $\sigma_t = 0.765\ 1/(\Omega - m)$ is the electromagnetic conductivity for the tissue. $\sigma_s = 0.87\ 1/(\Omega - m)$ is the electromagnetic conductivity for the skin. SAR is the specific absorption rate in W/kg which varies from 1.0 to 2.0 for the present study. At the present time most cell phones operate at a level of SAR=1.0 W/kg. US SAR limit for cell phones is at 1.6 W/kg for a volume that contains one gram of tissue which is absorbing the highest radiation. European SAR limit for cell phones is at 2.0 W/kg.

Using Equations 23-1 through 23-4 and above parameters, brain tissue temperature right inside the skull bone is calculated in a 60-minute cell phone voice call when the cell phone is placed next to the ear. Figure 23-1 shows the brain tissue temperature versus cell phone voice call time for two different environment temperatures, namely $T_{env} = 273.15\ K\ (0\ C)$ and $T_{env} = 323.15\ K\ (50\ C)$ at a specific absorption rate of 1.6 W/kg. With this SAR value $Q_{EM} = 773.8\ W/m^3$ using Equation 23-4. When the environment temperature is 0 C, the brain tissue temperature decreases by 2.78 C down to 34.22 C from the initial temperature of 37.0 C in 60 minutes. When the environment temperature is 50 C, the brain tissue temperature increases by 1.62 C up to 38.62 C from the initial temperature of 37.0 C in 60 minutes.

Sensitivity of brain tissue temperature changes with respect to increasing electromagnetic radiation energy exposure time as shown in Figure 23-2 for two different environment temperatures, namely $T_{env} = 273.15\ K\ (0\ C)$ and $T_{env} = 323.15\ K\ (50\ C)$ at the specific absorption rate of 1.6 W/kg. These sensitivities decrease in both cases, namely taking the absolute value of sensitivity in the $T_{env} = 273.15\ K\ (0\ C)$ case, as the electromagnetic radiation energy exposure time increases.

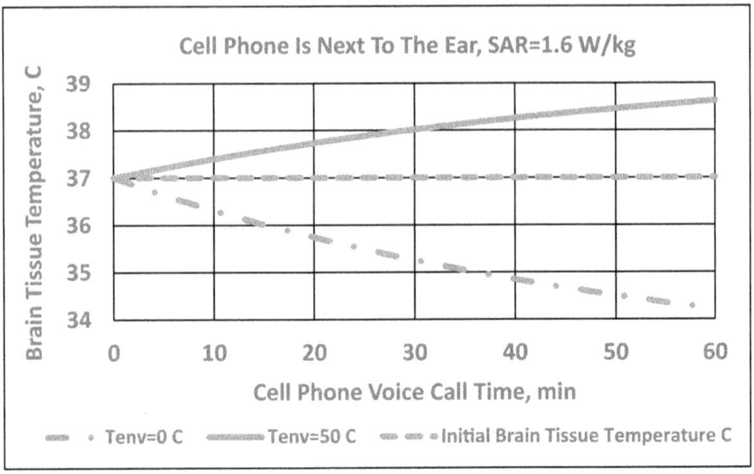

Figure 23-1 Brain tissue temperature versus cell phone voice call time for two different environment temperatures, namely T_{env} = 273.15 K (0 C) and T_{env} = 323.15 K (50 C) at specific absorption rate of 1.6 W/kg.

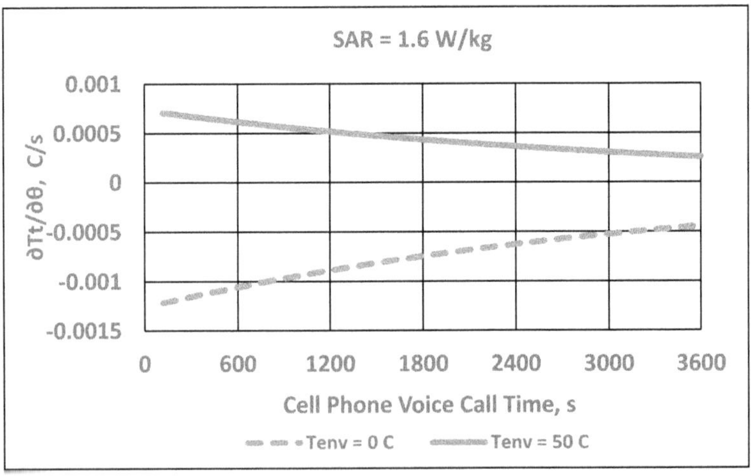

Figure 23-2 Sensitivity of brain tissue temperature change to increasing electromagnetic radiation energy exposure time for two different environment temperatures, namely T_{env} = 273.15 K (0 C) and T_{env} = 323.15 K (50 C) at specific absorption rate of 1.6 W/kg.

Brain tissue temperature changes in one hour exposure to electromagnetic radiation from a cell phone placed against the ear versus environmental temperatures from 0 C to 50 C are shown in Figure 23-3. The brain tissue temperature increase is almost linear to increasing environment temperature for the present SAR values of 1.0 to 2.0. Sensitivity of brain tissue temperature increase to environment's temperature increase is + 0.088 C per + 1.0 C. For the present calculations, the maximum brain tissue temperature drop is 2.96 C down to 34.22 C from the initial temperature of 37.0 C in 60 minutes when the environment temperature is 0 C and SAR=1.0 W/kg. For the present calculations, the maximum brain tissue temperature rise is 1.74 C up to 38.74 C from the initial temperature of 37.0 C in 60 minutes when the environment temperature is 50 C and SAR=2.0 W/kg.

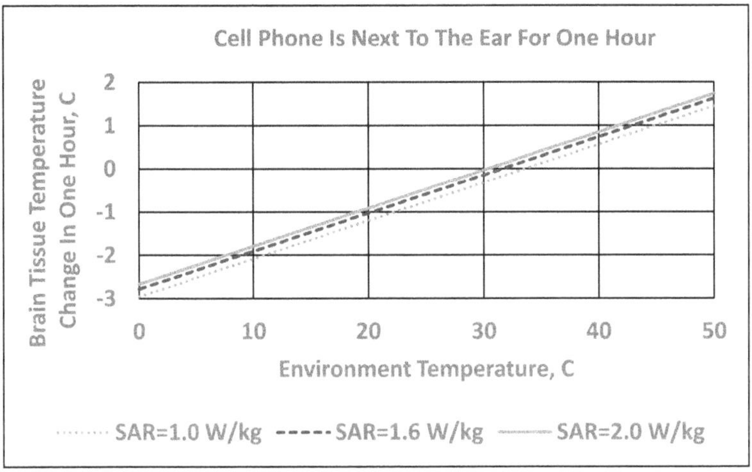

Figure 23-3 Brain tissue temperature change in one hour exposure to electromagnetic radiation from a cell phone placed against the ear versus environmental temperature.

Percentages of absolute values of heat transfer modes to total absolute value of heat transfer that are calculated with the present model are shown for 20 C environment temperature in Table 23-1. As the SAR values increase from 1.0 W/kg to 2.0 W/kg, temperature effects of electromagnetic radiation from a cell phone placed against the ear on the brain tissue increase as expected. As the SAR values increase from 1.0 W/kg to 2.0 W/kg, effects of blood perfusion heat transfer decrease since a constant blood perfusion rate is assumed in the present model. Convection heat transfer between the brain tissue and the outside air is the major heat transfer mode and it is over 50%. Radiation heat transfer mode is the minor heat transfer mode which is less than 2.5%.

Table 23-1 Percentages of absolute values of heat transfer modes to total absolute value of heat transfer that are calculated with the present model for T_{env} = 20 *C*.

SAR, W/kg	Convection %	Radiation %	Blood Perfusion %	Electromagnetic Radiation %
1.0	57.0	2.4	28.2	12.4
1.6	55.4	2.3	23.2	19.1
2.0	54.3	2.3	20.1	23.3

24

Cooling a Radar's Electronic Board

A radar, an acronym for **RA**dio **D**etecting **A**nd **R**anging, transmits from its transmitter short pulses of electromagnetic waves at radio and microwave frequencies, i.e. 30 kHz to 100 GHz, which may be reflected by objects on their path. Reflected electromagnetic waves come back to the receiving portion of a radar system to provide information about the shape and speed of the reflecting object such as airplanes, boats, missiles, speeding cars, etc. For example, radars in the 5 to 35 MHz electromagnetic wave transmission range are used to map currents in coastal waters. 2 to 4 GHz electromagnetic wave pulses are used for air traffic controllers. 8 to 12 GHz electromagnetic wave pulses are used for marine targets and missile guidance. 100 GHz electromagnetic wave pulses are used for high resolution meteorological observations and imaging. Rain particles also scatter portion of the electromagnetic wave pulses back to the receiver. This information tells where it is raining and how much precipitation there is. Doppler effect on these electromagnetic wave pulses can also determine the wind speed and direction of a rain storm.

Signal to noise ratio of a radar system enhances linearly with transmitted power and inversely with system temperature. Heat generating components of a radar system are the power supply, transmitter, receiver, digital processor and displays. Earlier radar system components and electronic boards had their temperature controlled by using air flow as a coolant. As transmitter powers increased, water and ethylene glycol were used for temperature control of electronic boards. Today's high power radars require further enhanced cooling such as two-phase flow systems which combine effects of forced convection cooling and evaporative cooling.

In this chapter, let us investigate temperature control of two electronic boards in a radar system using fully developed air flow in a rectangular duct as shown in Figure 24-1. Air cooling starts at the fully developed velocity profile region where the electronic boards' surfaces start.

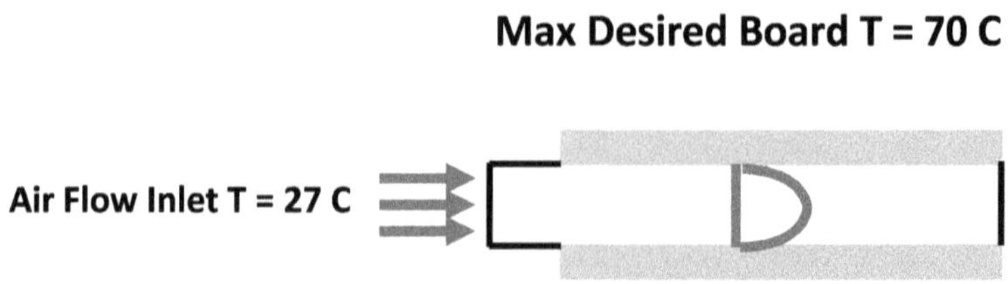

Figure 24-1 Cross-sectional view of two electronic boards being cooled by a fully developed turbulent air flow with an unheated air flow entry region in a rectangular duct to achieve a maximum board surface temperature of 70 C.

The electronic boards are 0.3 m long and 0.2 m wide. The air duct height is 0.05 m. Assuming constant and total heat flux, Q in W, from two electronic boards' surfaces to cooling air, the steady state heat transfer between two boards' surfaces and the fully developed turbulent air flow is given in Equation 24-1.

$$Q = \bar{h} \times A_s \times \left(T_s - T_{a-out}\right) \qquad 24\text{-}1$$

$\bar{h}$ is the average heat transfer coefficient in W/(m^2-K) for fully developed turbulent air flow in a rectangular duct given in Equation 24-2. $A_s = 0.12$ m^2 is the surface area of two electronic boards that are in contact with the cooling air. T_s is the maximum surface temperature requirement for normal functioning of the boards which is 70 C. $T_{a\text{-}out}$ is the average air outlet temperature from the cooling duct in C.

From the air cooling cross-section configuration given in Figure 24-1, the maximum surface temperature of electronic boards will occur at the outlet position of the air duct where the averaged air flow temperature, $T_{a\text{-}out}$, is the highest.

$$\overline{Nu}_{Dh} = \frac{\bar{h} \times D_h}{\nu_a} = 0.023 \times Re_{Dh}^{0.8} \times Pr_a^{0.4} \qquad 24\text{-}2$$

where $D_h = \frac{4 \times A_{cs}}{P} = \frac{4 \times 0.01}{0.5} = 0.08$ m for the present case. ν_a is the kinematic viscosity of air at average temperature of air inlet and outlet temperatures in m^2/s. $Re_{Dh} = \frac{\bar{U} \times D_h}{\nu_a}$ is the dimensionless Reynolds number for air flow which represents the ratio of inertia to viscous forces. $\bar{U}$ is the cross-section averaged air velocity in the cooling duct in m/s. $Pr_a = \frac{\nu_a}{\alpha_a}$ is the dimensionless Prandtl number which is the ratio of momentum to thermal diffusivity of air at average temperature of air inlet and outlet temperatures. Equation 24-2 applies for fully developed turbulent air flows when $Re_{Dh} \geq 10{,}000$, $0.6 \leq Pr_a$, and the air flow entry region is greater than $10 \times D_h$. See Ref. [13] for empirical correlations such as Equation 24-2 used in internal and fully developed turbulent flows.

If conservation of energy is applied to the fully developed velocity profile region of air flow with constant flow rate and with constant heat flux from the electronic boards' surfaces, Equation 24-3 is obtained.

$$Q = \dot{m} \times c_{p-a} \times \left(T_{a-out} - T_{a-in}\right) \qquad 24\text{-}3$$

$\dot{m}$ is the constant air mass flow rate through the rectangular duct in kg/s. c_{p-a} is the constant specific heat of air at constant pressure in J/(kg-K and it is taken at the average temperature of air inlet and outlet temperatures. T_{a-in} is the average and constant air temperature at the inlet of the rectangular duct and it is 27 C.

Cross-section averaged air velocity, $\bar{U}$, in the fully developed region of the rectangular duct is obtained from Equation 24-4.

$$\bar{U} = \frac{\dot{m}}{\rho_a \times A_{cs}} \qquad 24\text{-}4$$

ρ_a is the density of air in kg/m^3 in the rectangular duct and it is taken at the average temperature of air inlet and outlet temperatures. A_{cs} is the cross-sectional area of the air duct in m^2 and it is 0.01 m^2 for the present case.

Equations 24-1 through 24-4 are solved iteratively to determine the air mass flow rate required through the rectangular duct in order to have a maximum surface temperature of 70 C for the electronic boards for a given uniform and constant surface heat input. Required coolant air mass flow rate versus uniform and constant surface heat input is shown in Figure 24-2. Required coolant air mass flow rate increases in a nonlinear fashion as surface heat input increases. Sensitivity of a change in air mass flow rate to a change in surface heat input is shown in Figure 24-3. Sensitivity of changing air mass flow rate to changing surface heat input is 0.0006 kg/(W-s) at low values of surface heat input, i.e. 50 W. Sensitivity of changing air mass flow rate to changing surface heat input increases to 0.0012 kg/(W-s) around surface heat input of 1000 W.

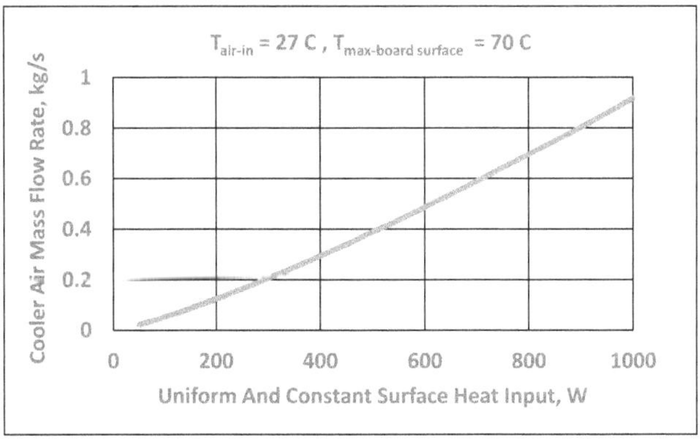

Figure 24-2 Required coolant air mass flow rate versus uniform and constant surface heat input.

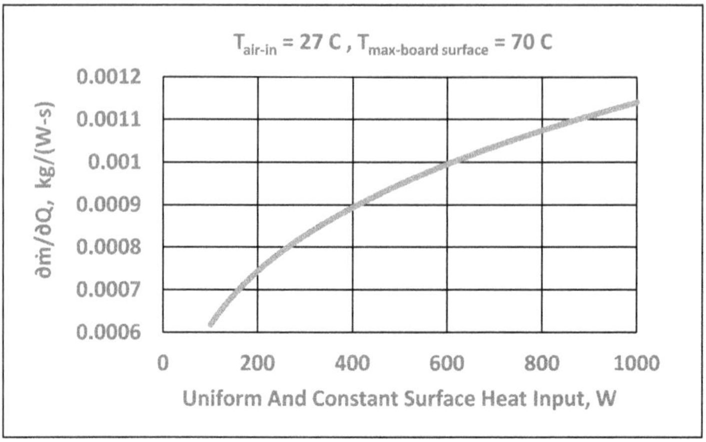

Figure 24-3 Sensitivity of air mass flow rate to surface heat input.

Fully developed turbulent air flow's Reynolds number versus uniform and constant surface heat input is shown in Figure 24-4. The Reynolds number is just about at 10,000 for a 50 W surface heat input and it increases to 400,000 when the surface heat input reaches 1000 W in order to keep the maximum surface temperature at 70 C for the electronic boards' surfaces.

It is interesting to note that throughout this case, a change in the average heat transfer coefficient for fully developed turbulent air flow in a rectangular duct, *d*h̄, to a change in the surface heat input, *dQ*, is approximately a constant at 0.2 1/(m^2-K).

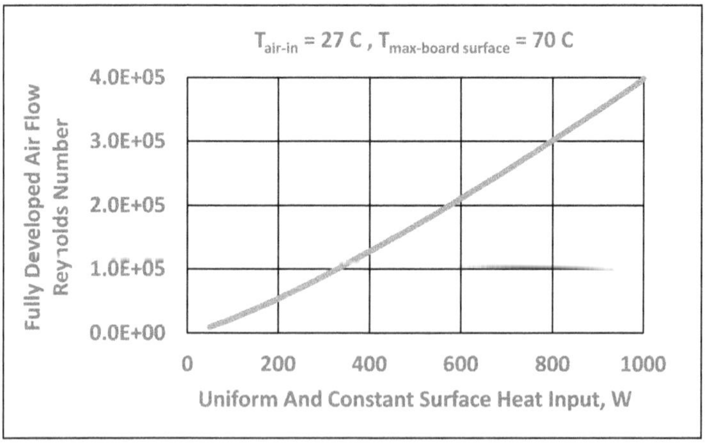

Figure 24-4 Fully developed turbulent air flow Reynolds number versus uniform and constant surface heat input.

References

1. ASHRAE (American Society of Heating, Refrigeration and Air-Conditioning), "ASHRAE Handbook", 2001.
2. Bernhard, G. and et al, "Measurements of Spectral Solar UV Irradiance in Tropical Australia", J. of Geophysical Research, V. 102, No. D7, 8719-8720, 1997.
3. Brode, H. L., "A Review of Nuclear Explosion Phenomena Pertinent to Protective Construction", 1964.
4. Cai, A. and et al, "Effects of Cell Phone Radiation on the Head", Computer Aided Engineering: Applications to Biomedical Processes, 2014.
5. Colorado State University, "Solar Irradiation Data: Fairbanks, Alaska", UV-B Monitoring and Research Program, 2019.
6. Duffie, J. A. and W. A. Beckman, "Solar Engineering of Thermal Processes", John Wiley & Sons, Inc., 1991.
7. Dumas, C. and G. S. Mittal, "Heat and Mass Transfer Properties of Pizza During Baking", International Journal of Food Properties, 2007.
8. FDA(Federal Drug Administration), "Criteria for Significant Risk Investigations of Magnetic Resonance Diagnostic Devices", fda.gov/media/71385, 2014.
9. Glasstone, S. and P. J. Dolan, " The Effects of Nuclear Weapons", US Dept. of Defense and US Dept. of Energy, Third Edition, 1977.
10. Health Physics Society, "Radiation Exposure from Medical Exams and Procedures", 2010.
11. ICNIRP(International Commission on Non-Ionizing Radiation Protection) Guidelines, "On Health Issues of Ultraviolet Tanning Appliances Used for Cosmetic Purposes", Health Physics, 84(1), 119-127, 2003.
12. ICNIRP(International Commission on Non-Ionizing Radiation Protection) Guidelines, "On Limits of Exposure to UV Radiation of Wavelengths Between 180 nm and 400 nm (Incoherent Optical Radiation)", Health Physics, 87(2), 171-186, 2004.
13. Incropera, F. P. and D. P. DeWitt, "Fundamentals of Heat and Mass Transfer", John Wiley & Sons, Inc, 1996.
14. International Energy Agency, www.iea-ebc.org/Data/publications/EBC Annex 12 therma and solar, 1987.
15. ISO/CIE Joint Standard, ""Global Solar Index" and "Erythema Action Spectrum and Standard Erythema Dose", ISO 17166, 1999.

16. Jakhrani, A. Q., "Estimation of Incident Solar Radiation on Tilted Surface by Different Empirical Models", International Journal of Scientific and Research Publications, V. 2, No. 12, 2012.
17. Klein, R. S. and et al, "Analysis of Compact Fluorescent Lights for Use by Patients with Photosensitive Conditions", Photochemistry and Photobiology, 10.1111, 2009.
18. Kodera, S. and et al, "Temperature Elevation in the Human Brain and Skin with Thermoregulation During Exposure to RF Energy", BioMedical Engineering, 2018.
19. Kreith, F., "Principles of Heat Transfer", International Textbook Company, 1965.
20. Lokmanhekim, M., "Procedure for Determining Heating and Cooling Loads for Computerizing Energy Calculations: Algorithms for Building Heat Transfer Subroutines", ASHRAE, 1971.
21. McKinley, A. F. and B. L. Diffey, "A Reference Action Spectrum for Ultraviolet Induced Erythema in Human Skin", Commission Internationale de l'Eclairage, 6(1), 17-22, 1999.
22. NASA, "Total Irradiance Monitor by NASA", https://atmospheres.gsfc.nasa.gov/climate/?section=136
23. NASA, "NASA Solar Systems", https://solarsystem.nasa.gov
24. NASA, "Surface Meteorology and Solar Energy: Methodology", www.ceoe.udel.edu/File Library/Research/Wind, 2004.
25. NREL(National Renewable Energy Laboratory), "Solar Radiation Data Manual for Flat Plate and Concentrating Collectors", 1961-1990.
26. P. Tierney and et al, "Nine out of 10 Sunbeds in England Emit Ultraviolet Radiation Levels That Exceed Current Safety Limits", British Journal of Dermatology, 2012.
27. Paolo V. and et al, "Guide Protecting Workers from UV Radiation", International Commission on Non-Ionizing Radiation Protection in Collaboration with International Labor Organization and World Health Organization, 2007.
28. Pitchai, K. and et al, "Coupled Electromagnetic and Heat Transfer Model for Microwave Heating in Domestic Owens", Journal of Food Engineering, 2012.
29. Struckmann, F., "Analysis of Flat Plate Solar Collector", Heat and Mass Transport, 2008.
30. Sutton, G. W. and A. Sherman, "Engineering Magnetohydrodynamics", McGraw-Hill, 1965.
31. WHO (World Health Organization), "Global Solar UV Index: A Practical Guide", WHO, 2002.
32. WHO(World Health Organization), "Electromagnetic Fields and Public Health: Mobile Phones", 2014.
33. Wiebelt, J. A., "Engineering Radiation Heat Transfer", Holt, Rinehart and Winson, Inc., 1966.

Index

A

Absolute values, 65
  of heat transfer modes, 197, 198
  of minimum sensitivity value, 72, 75
  radiation heat transfer, 179
  of sensitivity, 81, 87, 177, 178, 184
    for coated inner surfaces, 53, 54, 56
    surface emissivity, 148
    for uncoated inner surfaces, 53, 54, 56
Active system, 113
Actual average surface temperature, 11
Air burst, 62, 63
Air chamber environment, 150
Air velocity, 126, 127, 131
Algebraic equation, 35, 37
American Society of Heating, Refrigeration and Air-Conditioning Engineers (ASHRAE), 113
Aphelion, 10
ASHRAE. *See* American Society of Heating, Refrigeration and Air-Conditioning Engineers
Asymptote value, 154
Atmospheric transmittance, 186
August-Roche-Magnus formula, 135, 159

B

Bake temperature
  vs. time characteristics, 89
  vs. time specifications, 89
Base coat layer curing case, 97
Big Bang theory, 165
Biot number, 143, 149, 151, 157
Blackbody radiation, 9, 15, 18
  Planck's spectral distribution function for, 62
  vs. frequency, 23
Blast wave for air bursts, 59

C

Cell phones, 193–198
Changing emissivity, 120
Characteristic length, 151
Clearness index, 104–108
  values, 107–108
Coated inner surfaces, 57
  sensitivity for, 53–54, 56
Coating material emissivity, 55, 56, 58
Coffee *vs.* time, temperature of, 86, 87
Collector efficiency, 115–120
Communication satellites, 10

Conduction heat transfer, 133, 160
resistance, 85, 86
Configuration factor, 35–36, 38, 144, 176
cross-sectional geometric for, 36
using reciprocity expressions, 36
Conservation of energy, 90, 193, 201
Constant emissivities, 51
Constant pressure *vs.* water temperature, 167
Constant solar radiation, 51
Constant value, 134
Convection heat transfer, 47, 123, 133–135, 143, 194
coefficient, 47, 91, 116, 117, 124, 159, 194
laminar forced, 89–99
loss, 45
Convection mass transfer coefficient, 158
Cooling time
errors, 31
natural convection heat transfer *vs.*, 31, 154
sensitivity of, 31, 33, 155
special steel plate, 31, 33
and sphere ball, 152–153
of spherical steel ball, 155
Cosmic microwave, 165
Coupled electromagnetic wave equations, 165

D

Declination angle, 105
Dielectric loss factor, 168
degradation, 168, 169
Diffuse solar radiation, 106–109, 114
Diffusion of thermal energy, 166
Digital cellular networks, 193
Direct beam extraterrestrial radiation, 187–188
Direct beam solar radiation, 106, 107, 109, 114, 135

E

Earth
atmosphere, 10, 13, 104
gravitational acceleration, 28
radiation power absorbed by, 14
solar radiation on horizontal surface on, 104, 106
surface temperature, 12
UV radiation on, 69–75
EFF, 102
Electricity consumption demand, 104
Electrocoating, 89
Electromagnetic radiation, 11, 12, 35
effects on human brain tissue, 193–197
energy, 195
in medical diagnosis, 171–174
Electromagnetic waves, 9, 21, 77, 83
Doppler effect on, 199
by magnetron, 165
Maxwell's equations for, 193
radio frequency, 172
reflected, 199
Elemental surface, 41
Emissive power in Watts, 27
Emissivity of surface, 27

Empirical model, 104, 105
Energy transfer modes, 136, 137
Environment
for heavy clothed person, 131
human body and, 123
skin surface to, 124
temperature, 44, 126–130, 195
Equilibrium temperatures, 189
Erbs model, 106
Erythemal weighing function, 79
Evaporation rate, 134, 158
Explicit finite difference method, 29, 160
Exposure energy, ultraviolet radiation, 77
External radiant heat transfer resistance, 144
Extraterrestrial solar radiation, 104, 105, 107, 108, 111, 185–189
Extreme UVI exposure, 72

F

Fairbanks maximum sensitivities, 74, 75
Fairbanks solar irradiation data, 72
Finite difference method, 86
Finite surface areas, 35
Fluorescent lamps, 77
incandescent and, 80
UVR exposure for, 78
Forced convection heat transfer coefficient, 48
Free stream air velocity, 91–93
local heat transfer coefficient *vs.*, 94, 95, 98
Frequency space, 21–23

G

Gamma rays, 171–173
Glass heat transfer
for vertical double pane window, 47–58
Glazed flat plate solar radiation collector, 114, 115
Glazing gap, 49, 52
with air, 52
with argon, 54, 55
with krypton, 55–57
vertical double pane window, 56
Grashof number, 85, 86, 150, 159
Gravitational acceleration, 49
Gravitational constant, 160
Greenhouse effect
atmospheric, 14
human disturbance of, 13
Ground albedo, 106
Ground reflected solar radiation, 106

H

Heat energy, conversions solar radiation to, 113
Heat loss
convection and radiation, 123
sensible, 123–126, 128–131
through clothing, 124
Heat transfer
vs. baking time, 161, 162
coefficient, 187
average value of, 90
center of glass overall, 52, 58
convection, 124
forced convection, 48
vs. free stream air velocity, 94

local forced convection, 93–95, 97
natural convection, 48, 49, 114
radiation. *See* Radiation heat transfer
conduction, 133
convection, 133–135, 143
equations, 47
evaporation, 133
mechanisms, 133
Hemispherical absorptivity, 136
Hemispherical surface emissivity, 158
High UVI exposure, 72
Horizontal surfaces
equilibrium temperatures on, 189
under steady-state conditions, 186
zenith angle for, 186
Hottel-Whillier-Bliss equation, 114
Human thermal comfort, 123–132
without clothing, 127
Hypothalamus, 123

I

Ideal thermal radiator, 9
Incandescent lamps, 80
Incident solar radiation, 48, 114, 133
Independent variable, 115
Infrared radiation heat transfer, 50, 51, 84, 144, 149
baking pizza by, 157–164
Infrared radiation loss, 133
Insulating material, 113
Integration process, 69–70
Interfacial crust contact temperature, 162–164
Internal body temperature, 123
International Organization for Standardization (ISO) 9806:2017, 113
Ionizing electromagnetic radiation, 171, 172
natural resources of, 173
typical effective, 174
ISO Standard 17166, 69

K

Kinematic viscosity, 134, 135, 150
to thermal diffusivity, 49

L

Laminar forced convection heat transfer, 89–99
vehicle length averaged, 91–93
Least equilibrium temperature, 189
Linear function, 115
Liquid oxygen, 175–179
boiling point, 175
change in time for, 184
cryogenic spherical container, 175–176, 181
freezing point of, 175
heating times, 176
time for, 183
Local heat transfer coefficient, 93–95, 97
vs. free stream air velocity, 98
and vehicle length averaged heat transfer coefficient, 99

Low emissivity material, 52–54
Low UVI exposure, 72

M

Magnetic resonance imaging (MRI), 171–172
Mass transfer coefficient, 135, 159
Maximum exposure time
  sensitivity of, 78, 79, 81, 82
  vs. effective UVR exposures, 82
Maxwell's equations, 165, 166, 193
  for electromagnetic waves, 193
Medical diagnosis, electromagnetic radiation in, 171–174
Microwave radiation in cooking ovens, 165–169
  electromagnetic waves, 165
  heating time predictions in, 166
Millisievert, 173
Moderate UVI exposure, 72
MRI. *See* Magnetic resonance imaging

N

Natural convection heat transfer, 27, 32, 84–86
  coefficient, 48, 49, 114, 115, 190, 191
  vs. cooling time, 31
  resistance approaches, 154
Natural resources of ionizing radiation, 173
Non-ionizing electromagnetic radiation, 171, 193
Normal distribution function, 71, 72
Normalized thermal radiation power, 62, 63
Nuclear explosion, 59
  fireball, 62
  of thermal energy, 59
  thermal radiation fraction of, 63
  yield, 60, 64, 67
Nuclear fireball, 59
Nuclear medicine, 173
Numerical integration, 35
Nusselt number, 49, 134, 150

O

Operational variables, 99

P

Painted surface
  conservation of energy on, 90
  emissivity of, 90
  temperatures, 90, 93–95
  vehicle, 90
Passive system, 113
Percent emitted thermal radiation energy, 62, 63
Perihelion, 10
Photosphere, 9
Photovoltaic (PV) collector panels, 101–111
  in Anchorage, 103
  components on, 108, 111
  daily solar radiation on, 109–110
  efficiencies, 102
  electricity generation of, 101
  solar radiation on, 102–104
  total surface area of, 101
  yield, 102
Planck's spectral distribution function, 62

Planck's thermal radiation distribution function, 15–17
sensitivity of, 18–21
Prandtl number, 90, 134, 150, 159, 200
Pressure wave, 59
PV collector panels. *See* Photovoltaic collector panels

R

Radiant energy
diffuse, 187
fraction of, 144
surface emission, 187
Radiation heat transfer, 27, 32, 47, 83, 123
absolute values, 179
coefficient, 48, 50, 88, 124, 125, 143
infrared, 50, 51, 84, 144, 149
mode, 197
resistance, 85, 179
of spherical balls, 143
under steady-state conditions, 176, 182
in unsteady state, 157–164
value of, 183
Radiation power, 10–13
emissions, 16
estimates of, 13
Sun, 10, 16
Radio frequency
electromagnetic radiation, 195
electromagnetic waves, 172
Radiotracers, 173
Radio waves, 171
Rayleigh number, 28, 49, 150, 159
Reciprocity expressions, 36
view factor relationships with, 42
Rectangular thermal radiation heater, 41
Reflected electromagnetic waves, 199
Renewable energy sources, 12
Residual nuclear radiation, 59
Reynolds number, 90, 134, 202

S

SAR. *See* Specific absorption ratio
Saturation pressures, 159
for water vapor, 135
SED values. *See* Standard erythema dose values
Sensible heat loss, 123–126, 128–131
vs. air velocity, 131
with heavy clothing, 130
sensitivity of, 128, 132
Sensitivity
absolute values of, 81, 87, 177, 178, 184
of brain tissue temperature, 195–197
of changing air mass flow rate, 201
for coated inner surfaces, 53–54, 56
of coffee temperature, 87
vs. silver coating emissivities, 87
of collector efficiency, 119
of cooling time, 31, 33, 155
decrease, 67

environment temperature, 127, 128, 130, 131
of $F_{A2 \to dA1}$ *to radiant heater height,* 43, 44
of glazing gap width, 53
of heat loss, 137–141
of maximum exposure time, 78, 79, 81, 82
of maximum view factors, 46
of sensible heat loss, 128, 132
of thermal radiation, 18–21, 24–25
of thermal radiation exposure, 65, 66
of time
to sphere ball's radius, 147
to surface emissivity, 148
of time to cool, 152
for uncoated inner surfaces, 53–54, 56
of UV index, 72–74
Shape factors, 35
Shock wave, 59
Sievert (Sv), 173
Silver coating emissivity, 86, 87
60 Watt incandescent lamp, 79
Skin
electromagnetic conductivity for, 195
human, 81
temperature, 125
types, 80
Solar constant, 10, 186
Solar radiation
collector, 113, 114, 117
diffuse, 106–108
direct beam, 106, 107, 109
energy conversion, 101
extraterrestrial, 104–105, 107–108, 111
to heat energy conversions, 113
on horizontal surface on Earth, 104
incident, 48, 133
model, 110
optimization process, 101
on PV collector panel, 102–104
strikes, 47
to thermal energy, converting, 113–122
Solar spectral irradiation data, 69, 71, 72
integration process, 78
vs. wavelength, 70
Solar ultraviolet (UV) radiation, 69–75
exposure to, 75
irradiance measurement data, 70
Sound waves, 171
Special steel plates, 27–28
and environment, 28
surface temperature, 30, 32
thermal capacitance of, 28
Specific absorption ratio (SAR), 172, 195, 197
Spherical balls, 143–144
internal energy of, 144
and subscript, 150
surface emissivity, 156
temperature, 144, 146
vs. cooling times, 152
time dependent temperature, 151
Spherical cryogenic container, 175
Stainless steel bottle
inner bottle wall, 83, 84
outer bottle wall, 83, 84

silver coated surface, 85
surface area, 84
thermal conductivities, 85
of thermos bottle, 84
wall thickness, 83
Standard erythema dose (SED) values, 81
Standard MRI systems, 171
Steady-state conditions, 124, 182
heat transfer, 43–44, 47
horizontal surface under, 186
radiation heat transfer under, 176, 182
temperature, *vs.* time mapping, 89
Steel spherical balls, 149–156
Stefan-Boltzmann constant, 9, 15, 28, 35, 44, 50, 85, 124, 134, 144, 151, 158, 176, 194
Storing liquid oxygen, 181–184
Subscript, 150
Sun
emissivity, 79
photovoltaic collector panels, 101–111
radiation power, 10–11, 16, 22
surface temperature, 9
Surface emission radiation energy, 187
Surface emissivity, 148, 155, 176–177, 183
Surface temperatures, 45
blackbody, 22
frequency space and, 22
function, 27
special steel plates, 30, 32
Sun, 9
total hemispherical emissivity *vs.*, 28, 30
vs. wavelength, 15
wood bonfire, 18, 19
Swimming pools, energy balance for, 133–141

T

Temperature
of coffee *vs.* time, 86, 87
environment, 126–129, 195
without clothing, 129, 130
for spherical ball, 147
vs. time mapping, 89, 95, 99
Thermal energy, 113
balance, 123
of collector efficiency, 120, 121
converting solar radiation to, 113–122
diffusion of, 166
direct/indirect converter of, 113
nuclear explosion of, 59
Thermal infrared radiation, 11
Thermal radiation
configuration factor. *See* Configuration factor
distribution function, 14, 15
by blackbody, 17, 18
Planck, 15–17
sensitivities of, 18 21, 24–25
energy percent emitted, 62, 63
exposure, 63–65
heat transfer loss, 45
intensity of, 63

low level exposures *vs.* distance, 65
net exchange of, 35
Thermal radiation heater
to elemental surface, 43
rectangular and vertical, 41
sensitivity of $F_{A2 \to dA1}$ to, 43
steady-state, 43–44
view factor, 42
Thermal radiation power
blackbody, 62
emission of, 62
maximum value *vs.* nuclear explosion yield, 60–61
normalized, 62, 63
nuclear fireball, 59
standard measurement for, 59
Thermophysical properties, 51
Thermos bottle
heat transfer analysis, 86
heat transfer in, 83, 86
lid diameter of, 83
silver coated stainless steel surface, 85
stainless steel bodies of, 84
Time averaged efficiency, 114, 115
Time constant, 88
Time mapping, temperature *vs.*, 89, 95, 99
Time specifications, bake temperature *vs.*, 89
Total emissive thermal radiation power, 62
Total hemispherical emissivity
average, 27, 31–33
cooling temperatures, 29
steel surfaces, 27
vs. surface temperature, 28, 30, 31
temperature dependent, 31, 33
Townsville data, 70
Transient energy balance, 28
Transient non-linear differential equation, 29
Transmissivity, 84
Transmittance, 63, 64
Tungsten's emissivity, 79
Two-phase flow systems, 199

U

Ultraviolet radiation (UVR)
on Earth, 69–75
weighting function for, 70
emission, 77
exposure energy, 77
measurements of exposure, 81
standard erythema dose values, 81
Ultraviolet radiation index (UVI), 69, 78
exposure value, 72
vs. time of day, 72–74
for Townsville, 71
values and protection requirements, 71
Uncoated inner surfaces, 57
of panes, 58
sensitivity for, 53, 54, 56
Unexposed white skin, 80
Unsteady radiant heating, 143–145
Unsteady state
cooling process, 149
coupled heat transfer energy equation in, 166
heat transfer equation, 157
radiation heat transfer in, 157–164
sphere cooling, 150

V

Vacuum environment, infrared radiation in, 149
Vehicle length averaged heat transfer coefficient, 91
  primer coating layer of, 92
Vertical double pane windows
  central glazed portion, 52
  glass heat transfer in, 47–58
  glazing gap of, 56
  inner surfaces of, 53
  with uncoated inner surfaces, 54
  upper curve, 52
Vertical thermal radiation heater, 41
View factors, 35, 42
  reciprocity relationships, 42
Visible region radiation spectrum, 12
Volume of water, water boiling time *vs.*, 169

W

Water-air interface, 135
Water boiling time *vs.* volume of water, 169
Water density, 167
Water temperature
  constant pressure *vs.*, 167
  dielectric loss factor *vs.*, 168
  water density *vs.*, 167
Weighing functions, 70, 78
  erythemal, 79
Wind velocity, 138
Wood bonfire
  radiant power, 22
  surface temperatures, 18, 19
World energy consumption, 101
World Health Organization, 69
  solar UV index, 82

X

X-rays radiation, 171–173

Y

Yearly averaged daily solar radiation, 115
Yellow dwarf star, 14

Z

Zero sensitivity, 56

www.ingramcontent.com/pod-product-compliance
Ingram Content Group UK Ltd.
Pitfield, Milton Keynes, MK11 3LW, UK
UKHW020125290926
15471UKWH00003B/290